D0021779

Wind and Water

Wind and Water

JOHN TABAK, Ph.D.

For Rick Cardenas, an old friend, a good man.

WIND AND WATER

Facts On File, Inc.
An imprint of Infobase Publishing
132 West 31st Street
New York NY 10001

Library of Congress Cataloging-in-Publication Data

Tabak, John.
Wind and water / John Tabak.
p. cm. — (Energy and the environment)
Includes bibliographical references and index.
ISBN-13: 978-0-8160-7087-9
ISBN-10: 0-8160-7087-3
1. Renewable energy sources. I. Title.
TJ808.T33 2009
333.79′4—dc22 2008028247

Facts On File books are available at special discounts when purchased in bulk quantities for businesses, associations, institutions, or sales promotions. Please call our Special Sales Department in New York at (212) 967-8800 or (800) 322-8755.

You can find Facts On File on the World Wide Web at http://www.factsonfile.com

Text design by Erik Lindstrom
Illustrations by Accurate Art
Photo research by Elizabeth H. Oakes

Printed in the United States of America

Bang Hermitage 10 9 8 7 6 5 4 3 2

This book is printed on acid-free paper.

Contents

Preface

Nations around the world already require staggering amounts of energy for use in the transportation, manufacturing, heating and cooling, and electricity sectors, and energy requirements continue to increase as more people adopt more energy-intensive lifestyles. Meeting this ever-growing demand in a way that minimizes environmental disruption is one of the central problems of the 21st century. Proposed solutions are complex and fraught with unintended consequences.

The six-volume Energy and the Environment set is intended to provide an accessible and comprehensive examination of the history, technology, economics, science, and environmental and social implications, including issues of environmental justice, associated with the acquisition of energy and the production of power. Each volume describes one or more sources of energy and the technology needed to convert it to useful working energy. Considerable emphasis is placed on the science on which the technology is based, the limitations of each technology, the environmental implications of its use, questions of availability and cost, and the way that government policies and energy markets interact. All of these issues are essential to understanding energy. Each volume also includes an interview with a prominent person in the field addressed. Interview topics range from the scientific to the highly personal, and reveal additional and sometimes surprising facts and perspectives.

Nuclear Energy discusses the physics and technology of energy production, reactor design, nuclear safety, the relationship between commercial nuclear power and nuclear proliferation, and attempts by the United States to resolve the problem of nuclear waste disposal. It concludes by contrasting the nuclear policies of Germany, the United States, and France. Harold Denton, former director of the Office of Nuclear Reactor Regulation at the U.S. Nuclear Regulatory Commission, is interviewed about the commercial nuclear industry in the United States.

Biofuels describes the main fuels and the methods by which they are produced as well as their uses in the transportation and electricity-production sectors. It also describes the implications of large-scale biofuel use on the environment and on the economy with special consideration given to its effects on the price of food. The small-scale use of biofuels—for example, biofuel use as a form of recycling—are described in some detail, and the volume concludes with a discussion of some of the effects that government policies have had on the development of biofuel markets. This volume contains an interview with economist Dr. Amani Elobeid, a widely respected expert on ethanol, food security, trade policy, and the international sugar markets. She shares her thoughts on ethanol markets and their effects on the price of food.

Coal and Oil describes the history of these sources of energy. The technology of coal and oil—that is, the mining of coal and the drilling for oil as well as the processing of coal and the refining of oil—are discussed in detail, as are the methods by which these primary energy sources are converted into useful working energy. Special attention is given to the environmental effects, both local and global, associated with their use and the relationships that have developed between governments and industries in the coal and oil sectors. The volume contains an interview with Charlene Marshall, member of the West Virginia House of Delegates and vice chair of the Select Committee on Mine Safety, about some of the personal costs of the nation's dependence on coal.

Natural Gas and Hydrogen describes the technology and scale of the infrastructure that have evolved to produce, transport, and consume natural gas. It emphasizes the business of natural gas production and the energy futures markets that have evolved as vehicles for both speculation and risk management. Hydrogen, a fuel that continues to attract a great deal of attention and research, is also described. The book focuses on possible advantages to the adoption of hydrogen as well as the barriers that have so far prevented large-scale fuel-switching. This volume contains an interview with Dr. Ray Boswell of the U.S. Department of Energy's National Energy Technology Laboratory about his work in identifying and characterizing methane hydrate reserves, certainly one of the most promising fields of energy research today.

Wind and Water describes conventional hydropower, now-conventional wind power, and newer technologies (with less certain futures) that are being introduced to harness the power of ocean currents, ocean waves, and the temperature difference between the upper and lower layers of the ocean. The strengths and limitations of each technology are discussed at some length, as are mathematical models that describe the maximum amount of energy that can be harnessed by such devices. This volume contains an interview with Dr. Stan Bull, former associate director for science and technology at the National Renewable Energy Laboratory, in which he shares his views about how scientific research is (or should be) managed, nurtured, and evaluated.

Solar and Geothermal Energy describes two of the least objectionable means by which electricity is generated today. In addition to describing the nature of solar and geothermal energy and the processes by which these sources of energy can be harnessed, it details how they are used in practice to supply electricity to the power markets. In particular, the reader is introduced to the difference between base load and peak power and some of the practical differences between harnessing an intermittent energy source (solar) and a source that can work virtually continuously (geothermal). Each section also contains a discussion of some of the ways that governmental policies have been used to encourage the growth of these sectors of the energy markets. The interview in this volume is with John Farison, director of Process Engineering for Calpine Corporation at the Geysers Geothermal Field, one of the world's largest and most productive geothermal facilities, about some of the challenges of running and maintaining output at the facility.

Energy and the Environment is an accessible and comprehensive introduction to the science, economics, technology, and environmental and societal consequences of large-scale energy production and consumption. Photographs, graphs, and line art accompany the text. While each volume stands alone, the set can also be used as a reference work in a multidisciplinary science curriculum.

Acknowledgments

The author extends special thanks to Stan Bull, Associate Director for Science and Technology for the National Renewable Energy Laboratory, for sharing his time and considerable insights. Also important to the preparation of this work were Ken McDonnell, Senior Media Relations Specialist with ISO New England; Elizabeth Oakes, for her creativity in assembling the photography; and Frank Darmstadt, executive editor.

Introduction

Energy is one of the fundamental problems of the 21st century, although there is no lack of it. There are enormous reserves of energy in the winds, the tides, and in the temperature difference between the upper and lower regions of the oceans. It is an oft-repeated claim that each of these sources of energy is so abundant that if the energy of any one of them were converted into electrical energy, it would satisfy the electricity demands of the entire world many times over. This claim is true, but the implication that one could under any circumstance convert more than a tiny fraction of any one of these energy sources into electricity is false. There is no way to effect such a conversion—not now, not ever. Understanding both the promise and the limitations of these energy sources is one of the principal aims of this book.

All of the technologies in this book have benefited from significant government subsidies. *Conventional hydroelectric power* projects with their enormous dams and their enormous reservoirs cannot be built without significant government funding and other forms of assistance. Once built, they are often operated in ways very different from those of for-profit generating stations and often for good reasons. As for the other technologies discussed in this book, they also require significant subsidies to construct—and so far they have also required significant subsidies to operate. Understanding how these power producers benefit from government support, the economics of operating these types of generating stations, and the ways that they contribute to the power pool is an important part of understanding their potential, and that is the other main objective of *Wind and Water.*

Chapters 1 through 3 describe conventional hydroelectric power. Measured by the size of its contribution, conventional hydroelectric power is the most important technology discussed in this volume, but it is a mature technology, highly refined and not subject to much additional improvement. In most nations, most of the commercially valuable sites have already been developed. Significant changes in this sector are unlikely.

The second part of the book consists of chapters 4 through 7, and describes wave energy converters, *tidal mills,* ocean thermal energy converters, and *tidal barrages.* The first three of these technologies are rapidly evolving. They have, therefore, attracted a great deal of media attention, but so far they have generated relatively little power. Tidal barrages, an old idea, is being reexamined as those nations with suitable sites reconsider the technology in light of higher fossil fuel prices, increased demand for electricity, and the desire to generate more electricity with fewer emissions. All these technologies are heavily dependent on government support for their construction and operation. They are, therefore, economically uncompetitive. But given help, they may become competitive. Some believe that if governments intervene aggressively enough on behalf of these technologies, they can rapidly create markets for them.

The role of government in developing markets for power-production technologies is examined in chapter 7, with special reference to wind power, the most mature of these power technologies. Pro-wind policies in Denmark, Germany, and the United States are described and compared. This chapter also contains an interview with Dr. Stan Bull, associate director for science and technology for the National Renewable Energy Laboratory, the nation's premier site for research into renewable energy, about the challenges involved in establishing and maintaining high-quality government-funded research programs.

The last part of the book, consisting of chapters 8 through 10, describes the technology of wind *turbines,* the ways that wind turbines can be harnessed to contribute to the power supply, and the politics of wind power, especially the role of economic class in the siting of wind farms. All of these issues have so far played an important role in the development of this rapidly growing segment of the power market.

From an environmental viewpoint, all of the technologies described in this volume are relatively benign. The environmental impacts of even the most disruptive of them—namely, the very large hydroelectric projects—are generally local in nature. In particular, none of these technologies affect the global climate. But for a power-production technology to be adopted, it is not enough that it not pollute. It must also produce power at the right time at the right price and in sufficient quantities to meet demand.

Wind and Water seeks to bring the engineering, economic, environmental, and public policy aspects of these technologies together to provide an overview of their place in the power sector and thereby contribute to the ongoing debate about the ways that electricity should (or even can) be generated in the future.

PART I

Hydroelectric Power

Waterpower: A Brief History

Conventional hydroelectric power, the technology that converts the energy of moving water into electrical energy, depends upon a complex union of ideas and technologies. Some of these ideas and technologies were pioneered thousands of years ago; some are of much more recent origin. What is certain is that no one person invented the technology of hydroelectric power. Instead, hydropower evolved in response to numerous and sometimes conflicting engineering, economic, and societal requirements. To appreciate how modern hydroelectric plants work, it helps to know something of their history. That is the purpose of this chapter.

DESIGNS FROM ANTIQUITY

The earliest devices to convert the energy of moving water into work were waterwheels. The ancient Greeks used a type of waterwheel to grind grain. The earliest extant reference to these waterwheels is in

Waterwheel on the Orontes River, photographed between 1898 and 1914. The size of the project indicates the importance that the people of the time attached to this device. *(Library of Congress)*

a poem by the Greek poet, Antipater of Thesselonica (ca. 85 B.C.E.), who wrote the following:

> Cease grinding, ye women who toil at the mill; sleep late, even if the crowing cocks announce the dawn. For Demeter has ordered the Nymphs to perform the work of your hands, and they, leaping down on the top of the wheel, turn its axle, which with its revolving spokes, turns the heavy Nysarian millstones. We taste again the joys of the primitive life, learning to feast on the products of Demeter without labor.

It is believed that the type of waterwheel Antipater praises was placed horizontally in the water with only part of the wheel in the flowing stream. The vertical axle extended upward, where the grinding device was attached to the axle. The water, pushing against part of the wheel, provided the force necessary to do the grinding.

The ancient Romans, Chinese, and Japanese all used waterwheels of various types. Some wheels were oriented vertically, some horizontally; sometimes gears were used. The exact details of construction were determined by the speed with which the water flowed, the volume of water pushing against the wheel per unit time, and the technical sophistication of the designers. What all waterwheels have in common is that they convert linear motion into rotary motion. In particular, they use moving water to turn a shaft. This is critical because a rotating shaft can drive any piece of machinery.

Waterwheel design evolved over time, but most of the early records regarding their construction and use have been lost. It is known, for example, that the Romans constructed an enormous complex at Barbegal, located near Arles, France, of 16 seven-foot (2 m) waterwheels. These waterwheels were built in pairs, and the pairs were constructed at descending levels along a hillside. An aqueduct brought water to the top of the complex. The water flowed downward from the aqueduct turning each pair of waterwheels in sequence. The turning wheels provided the power necessary to grind grain. Estimates of the capacity of this enormous facility vary widely. Various authorities estimate that the Romans ground sufficient grain at this one facility to provide food for a population consisting of from 12,000 to 80,000 people. Whatever the number, it is certain that its construction required a large commitment of resources, that its maintenance involved numerous skilled individuals, and that it served a large community. The facility was a valuable resource.

Waterwheels remained important throughout the Middle Ages. The *Domesday Book*, the main source of information on 11th-century England, reveals that there were more than 6,000 water-powered mills in England when the book was compiled. All of these mills were, apparently, used to grind grain.

But a waterwheel turns only when there is sufficient water flowing past it. Seasonal variations in rainfall as well as local weather conditions affect the volume of water flowing past any mill. Whenever the water dries up, so does the machine's output. One way to introduce more predictability into watermill performance involves coupling the watermill to a dam. The dam enables the mill owner to store water when it is plentiful and the flow is abundant, and release water whenever it is needed. While this does not guarantee that there will always be sufficient water—a long drought can exhaust the storage capacity of the reservoir—it does enable the waterwheel owners to maintain production during periods of short-term water scarcity.

In addition to providing water storage, dams often serve to increase waterwheel *efficiency.* By raising the water level, one can increase the force with which the water impinges on the blades of the waterwheel, with the result that more work can be extracted per unit of water. Ancient waterwheels (and modern water turbines) work *between* water levels. If the water level upstream of the wheel and the water downstream of the wheel are equal, no water will flow; the wheel will not turn. The larger the difference in height between the upstream and downstream levels, the faster the water flows and the more work can be accomplished per unit volume of water. Increasing this difference in height is one way that ancient and modern designers extract more work per unit of water.

As with waterwheels, early records of dams are also scarce. It is, however, known that dams have been constructed since the beginning of recorded history and perhaps earlier. In particular, it seems likely that dam technology is at least as old as waterwheel technology. Dams were, for example, constructed in ancient Egypt. Egyptian records describe a dam constructed across the Nile in 2900 B.C.E. That dam is gone, but another, built in Syria on the Orontes

River in 1300 B.C.E., remains in use. Early dams were constructed for the purpose of storing water, which could later be used to furnish drinking water or water for irrigation. The decision by designers to couple one or more waterwheels to a dam was an important step forward in the development of waterpower, but when this first occurred is unknown.

THE INDUSTRIAL REVOLUTION AND MICHAEL FARADAY

The Industrial Revolution is the term used to describe the Western transformation from an agrarian economy to an industrial one. Eighteenth-century Britain was the first to experience the Industrial Revolution, but it spread rapidly—first to Belgium and France and soon thereafter to the United States. In Britain coal provided the energy needed to power the transformation. While coal was important everywhere, in the United States waterpower was also an important power source.

Much of the eastern United States enjoys annual rainfall levels of about 40 inches (1 m) per year. The topography of the eastern United States is such that water flows rapidly to the sea. This is an area that is ideally suited to the practice of harnessing water to do mechanical work. Lowell, Massachusetts, was the site of the first successful large-scale development of waterpower for manufacturing purposes. It began during the 1820s when Boston merchants began to build the nation's first planned industrial city along the Pawtucket Falls section of the Merrimack River. Along the Falls and in less than a mile (1.6 km), the Merrimack River descends 32 feet (9.6 m). To harness this drop in water level, city planners constructed nearly six miles (10 km) of canals, and they constructed a dam at the head of the falls to divert water through the canals. The flowing water turned waterwheels and their more modern analogues, water turbines, and the spinning shafts of these devices were connected to the machinery in the textile mills. There was no electricity in these early mills; this was a purely mechanical use of hydropower. Mill machinery was driven by often-complicated systems of belts and pulleys that were harnessed to the spinning shafts that were attached to the turbines that were driven by the flowing waters of the Merrimack River. This was not hydroelectric but rather hydromechanical power. The mills' machines could not be located farther from the water than one could stretch a series of drive belts. Despite this severe limitation, the mills at Lowell produced enormous wealth and were considered by the people of the era a tremendous success.

Within 20 years from the time that work at Lowell had begun, multistory mills lined the canals with a combined capacity of 10,000 horsepower. Lowell had gone from a tiny village to the second largest city in the state. The Lowell experiment served as a model for other industrial cities located elsewhere in the nation. Holyoke, Massachusetts, Nashua, New Hampshire, Saco, Maine, Augusta, Georgia, and Cohoes, New York, were some of the cities whose development was patterned on that of Lowell. The power of water to drive machinery had been demonstrated in Lowell as never before.

The next step in the development of hydroelectric power grew out of the work of the British scientist Michael Faraday (1791–1867). Faraday, who was originally trained as a bookbinder and who began his scientific work in the field of chemistry, is, today, best remembered for his discoveries about the interplay between electricity and magnetism. Endowed with an almost numbing patience, a lively imagination, and a profound intellect, Faraday was also an accomplished inventor, and during his decades-long series of experiments with electrical and magnetic phenomena he created a remarkable collection of devices to illustrate his scientific ideas and sometimes to discover new ones.

Here is one of Faraday's most important discoveries: By spinning a copper disk between the poles of a magnet and placing one

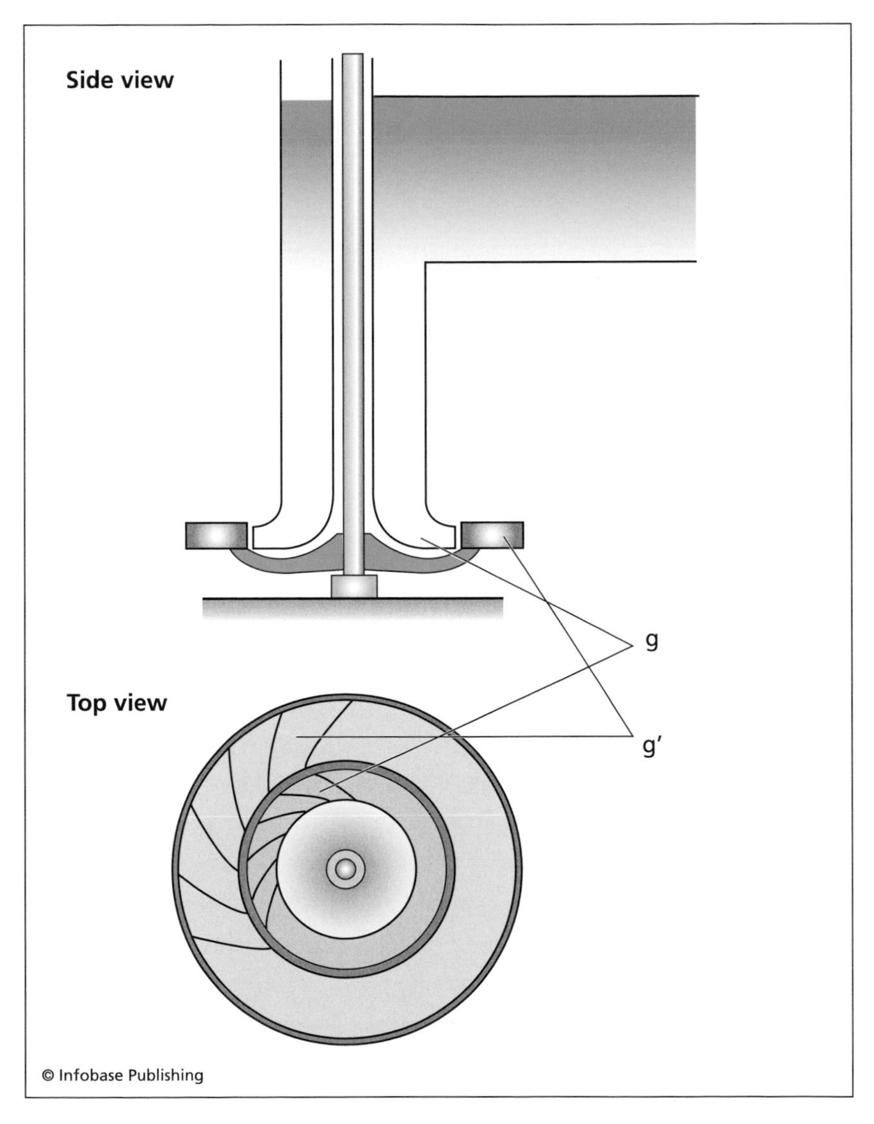

A cutaway diagram of an early type of water turbine called a Fourneyron turbine. Water flows downward along the shaft and is then directed outward by stationary vanes at *g*. As the water flows away from the shaft, it pushes against so-called buckets at *g′*. The buckets are attached to the rim of a wheel, causing the wheel and shaft assembly to spin. The shaft transmits power upward where it is harnessed to do work.

Michael Faraday. His generator converted mechanical energy into electrical energy. *(University of Aberdeen, Department of Physics)*

end of a circuit in contact with the rim of the disk and the other end in contact with its center, he discovered that he could create a steady electrical current. This was the first electric *generator.* By turning the shaft on which the disk was mounted, he produced an electrical current for as long as he kept the shaft spinning. This was extremely important because in contrast with the already-invented battery, Faraday's power source functioned indefinitely; it never ran out of electricity; it never went dead. He had turned the generation of electricity into a mechanical—as opposed to a chemical—problem. He demonstrated that to produce a continuous supply of electricity, one need only harness a continuous power source to the shaft of the generator.

But Faraday was not satisfied. He tried replacing the disk with wires and again with a metal sphere. He investigated different geometrical configurations for his generator and attempted to compare each to the other in order to determine its efficiency in producing current. He knew that given sufficient time and funding he could have built larger, more powerful generators, but he did not do so. His interests lay elsewhere. Faraday sought to discover broad physical principles; he sought to reveal new laws of nature. In effect, Faraday left the following two problems for future generations of engineers to solve: (1) Increase the size and efficiency of the generator in order to produce greater electrical power; and (2) Harness power sources capable of steadily spinning the shafts of these ever more massive generators. The engineers succeeded. They continue to succeed, and the result is an increasingly abundant supply of electricity.

Although the technology has changed a great deal since Faraday constructed the first generator, the concept has not. Broadly speaking, a modern power plant generator, the piece of technology that actually produces electricity, has two pieces, a *rotor* and a stator. The cylindrical rotor fits inside the hollow (and stationary) stator the way

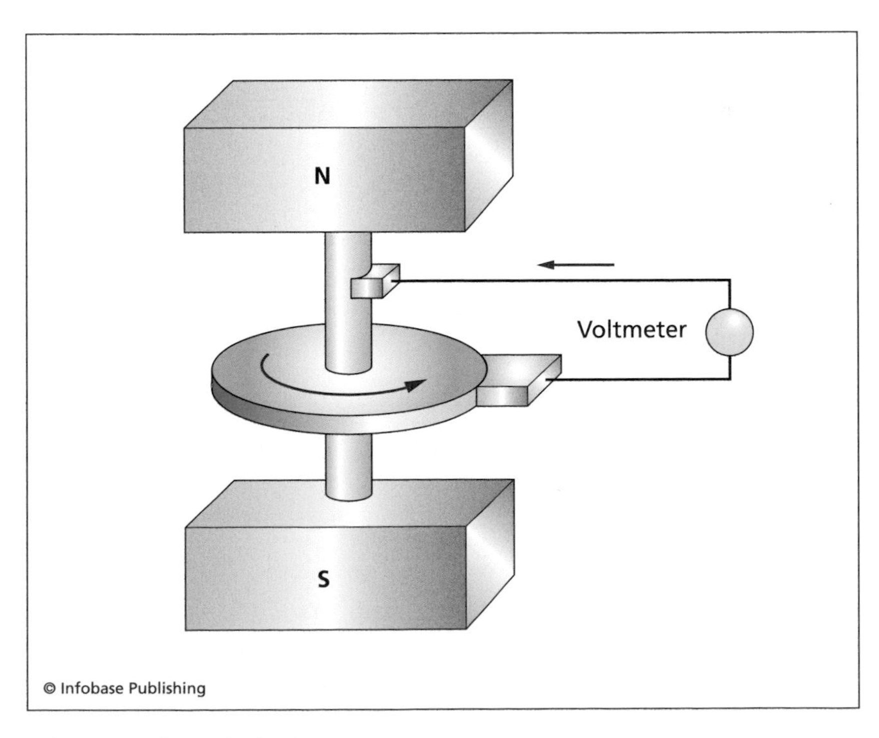

Schematic of Faraday's generator

an ink cartridge fits inside a pen, and then, as its name implies, the rotor rotates. For as long as it rotates within the stator an electric current emanates from the generator: The more powerful the source used to spin the rotor, the larger and more powerful the generator that can be used.

To be clear, generators convert mechanical energy into electrical energy. They do not create energy. One cannot extract more electrical energy from the generator than the amount of mechanical energy that one uses to spin the rotor. In particular, the greater the "load," or electrical demand, placed on the generator, the more force is required to turn the rotor. These simple-sounding facts explain why generators come in different sizes and produce differing amounts of electricity. In the case of water turbines, for example, moving water pushes directly against the blades of the turbine, which converts the linear motion of the water into the rotary motion necessary to spin the rotor. The harder the water pushes, the more electrical energy can be generated.

Faraday's invention made it possible, in theory, to harness the spinning shafts of turbines and waterwheels to generators and produce electricity, but his work did not immediately lead to hydroelectric power, because initially there was no demand for it. There were no electrical appliances.

CREATING ELECTRICITY DEMAND AND SUPPLY

Sir Humphry Davy (1778–1829), British chemist and teacher of Michael Faraday, was a prominent scientist in his own right and received numerous honors during his life for his discoveries. One discovery made by Davy was the arc lamp. He found that by placing two pieces of carbon close together—he used charcoal—and passing a current between them an arc of brilliant light formed. The light emanates from the electric arc that forms across the gap between the two pieces of carbon and from the ends of the pieces of carbon. In 1807, Davy used charcoal sticks and an enormous battery to produce the light. The invention had no practical value because no practical source of inexpensive electricity yet existed, but Davy's discovery did inspire a number of engineers to begin the process of improving the technology. Several important modifications were made by a number of engineers over the next several decades with the result that arc lighting was sometimes used to illuminate public places on special occasions. (Arc lighting is too bright and too harsh to use as indoor illumination.) The technology was not, however, widely adopted, because batteries were simply too expensive to use as a regular source of power, and no other practical source of power was yet available.

Another source of light, the incandescent lightbulb, was invented by the American inventor Thomas Edison (1847–1931) in 1879. To make money from his invention Edison had to overcome several obstacles. In particular, he had to construct an electric power infrastructure to generate electricity and transmit it to the consumers who wanted to use his bulbs to illuminate their homes and businesses. Edison's incandescent lamp, which was well-suited for indoor illumination, and the arc lamp, which was well-suited for outdoor illumination, created a demand for electricity. Initially, however, hydropower was poorly suited to supply this market.

To appreciate how the demand for electricity was met, keep in mind that electricity can be delivered as *direct current* (DC) or *alternating current* (AC). As their names imply, direct current flows along a wire in one direction only, much as water flows within a pipe. By contrast, alternating current regularly reverses direction. A cycle of AC current consists of flow in one direction and then in the other. Today, in North America it is standard practice to deliver AC electricity to homes at 60 cycles per second. This is called the frequency of the current. Edison, however, provided DC current, the same type of current generated by batteries, to his customers, and at first Edison's choice seemed reasonable. There were only a few customers, and the generating stations were built near where the demand was. The problem with DC current—a problem that has since been overcome—was that it was poorly suited for transmission over longer distances. The losses were too great—that is, as the electricity traveled along the power line some of the electrical energy was converted into thermal energy due to the resistance of the wire to the flow of electricity. The farther the customer was located from the generator, the less electrical power he or she received. Transmitting DC current using the technology of Edison's time was akin to sending water through leaky pipes. The problem was a serious one, because most major sources of hydroelectric power—Niagara Falls is the most famous example—were located too far from most potential consumers to supply them with DC electricity using the technology available at the time. At this point in the history of hydroelectric power, there were electrical appliances, which created a demand for electricity, and simple generators and turbines, which provided a supply. But there was not yet the technology available to connect electricity suppliers with electricity consumers if they were located more than a few miles apart.

While Edison was busy promoting the use of DC current, another approach to power generation and transmission was being proposed. This second approach used AC current. The backers of AC proposed a somewhat more complicated method of distributing power that depended on devices called *transformers*. Electrical transformers change the *voltage* of AC current. One feeds electricity into one side of the transformer at a given voltage and, depending on how the transformer is built, the AC current emerges from the other side with the same frequency but with a voltage that has a different but very specific relationship to the original voltage. One can construct transformers that increase the voltage by a factor of 2, 10, 100, or any other ratio. These are called *step-up transformers*. Similarly, one can use transformers to lower the voltage by any predetermined factor. These voltage-lowering transformers are called *step-down transform-*

ers. This is important because when AC electricity is transmitted along electrical lines at high voltages the losses that plagued Edison's low voltage DC system are very much reduced—the higher the voltage at which the electricity is transmitted, the lower the losses—and when the electricity nears its destination, the voltage can be lowered by passing the electricity through a step-down transformer. This was the technology that would make it possible to connect hydroelectric power plants with consumers located in distant cities.

More than anyone else, the American engineer and businessman George Westinghouse (1846–1914) was responsible for the technology used to bridge U.S. consumers with hydroelectric suppliers. He was not the first person with the idea of using AC power. The idea was pioneered in Europe. Westinghouse had, in fact, bought the American rights to patents that had already been obtained in Europe by the French and English team of Lucien Gaulard and John D. Gibbs. Westinghouse improved on the technology, but he was far from the originator of it.

What made the technology championed by Westinghouse so important from the outset is that most significant potential sources of hydroelectric power were not located near major markets. This was especially true in the western regions of the United States. There, rapidly flowing streams are often located in the mountains, but the population centers of the time were largely confined to the coast. Portland, Oregon, for example, was one of the earliest cities to receive electric service. In 1889, the Willamette Falls Electric Company began operation of a new facility along the Willamette River in Oregon City. The AC electricity produced in Oregon City was sent to Portland, located 13 miles (21 km) away, along a 4,000-volt transmission line. In Portland it was stepped down to 50 volts. Again, it was Westinghouse's AC technology that provided the bridge between U.S. consumers and producers of electricity.

(continued on page 18)

Niagara Falls

Today most people know Niagara Falls as a tourist attraction, but during the 19th century, when an energy-poor United States was seeking the sources of energy it would need to develop, Niagara Falls was perceived primarily as a potential energy source. Attempts to begin utilizing the power of falling water at Niagara began in the 1820s when two canals were built along the rapids that exist just above the falls. The energy of the water that flowed through these canals was used to power mills, but utilizing the main source of energy, the 180-foot (55-m) drop that occurs at the falls themselves, was beyond the technology of the times.

Nineteenth-century development at Niagara took two paths. Canal builders continued to expand the canal system, first along the rapids and then below the falls along Niagara Gorge, which was soon lined with an unbroken row of mills. But the canal systems did not tap the real power of the falls themselves. Instead, they made use of the relatively small drop in height along the river immediately before and immediately after the falls. By the 1880s, the turbines and generators needed to produce the power and many of the electrical appliances that would consume it all existed. In 1886, the Niagara River Hydraulic Tunnel Power and Sewer Company was organized to dig deep tunnels that would divert water to the nearby city of Niagara Falls. The tunnels would produce *hydraulic heads* of 79 to 125 feet (24.1 to 38.1 m). (The hydraulic head is defined as the difference in height between the surface of the water upstream and the surface of the water downstream.)

Financial and legal problems prevented the success of the plan, and the Cataract Construction Company purchased Niagara River Hydraulic. Cataract Construction would be responsible for the final tunneling. But this was a project designed to produce profit as well as power. The central financial barrier to the project was that the city of Niagara Falls, with a population of only 20,000, was simply too small a market for such a large and expensive-to-develop power source. The nearest big market was Buffalo, New York. But Buffalo was located 26 miles (42 km) away. Bridging such a long distance was a new type of technical challenge.

Cataract solicited ideas for power transmission schemes from many sources. Thomas Edison, America's premier inventor, was one of the first to be consulted by Cataract, and, not surprisingly, he lobbied for installing a DC transmission line, which was the technology in which he had invested. The difficulties of transmitting DC power 26 miles were, however, apparent to everyone concerned. Others advocated using compressed air to transmit power. Compressed-air technology was receiving some attention in Europe at the time.

The technology that eventually found favor was the one championed by George Westinghouse: Electricity would be generated at one voltage, the voltage would be increased (to 11,000 volts) for transmission to Buffalo and then stepped-down to 110 volts for small motors, arc lamps, and incandescent lighting, 100 or 250 volts for streetcar lines, 240 to 2,000 volts for various industrial applications, and so on. In 1895, the facility at Niagara began serving local customers. In 1896, the system operators began transmitting power to Buffalo. The generating station installed at

(continues)

Niagara Falls, mills on the American shore, ca. 1900. A century ago one of America's most famous tourist attractions was an industrial site. *(Library of Congress)*

(continued)

Niagara Falls was a financial success. It not only met demand, it created it. Soon energy-hungry industries—two famous examples of which are ALCOA, then known as the Pittsburgh Reduction Company, and Union Carbide—established large facilities in the vicinity of Niagara Falls to take advantage of the power produced there. The Niagara Falls project proved the viability of large-scale hydroelectric generation and large-scale electricity transmission, and set the pattern for the many large hydroelectric projects that followed.

(continued from page 15)

By the beginning of the 20th century, all the necessary subsystems for the production, transmission, and consumption of hydroelectric power were in place. The hydroelectric power production and distribution systems of the time were, by today's standards, extremely primitive, but they had, nevertheless, been developed to a high enough degree to make large-scale projects practicable. Numerous innovations were on the way, but the hydropower resources of the United States, Canada, and other countries with a sufficient technological base were now open for development. Electricity demand surged in the United States during the early decades of the 20th century as did production *capacity,* and for a while, hydroelectric power kept pace. During the 1920s, about 40 percent of the nation's electrical needs were met with hydropower.

Theory and Practice

Hydroelectric power occupies a special niche in the electricity markets. That niche has evolved as the electricity markets have evolved. The purpose of this chapter is to understand some of the fundamental principles that govern the operation of hydroelectric generating stations and how these principles have affected the role of hydroelectric plants in the electric power markets.

The purpose of a hydroelectric facility is to convert the power of moving water into electrical power. Power is defined as energy per unit time, so statements about power are statements about the rate at which energy is supplied. Plant efficiency is the relationship between the power provided to the plant (in the form of moving water) versus the power manufactured by the plant (in the form of electricity). When thinking about "power in" (in the form of water) versus "power out" (in the form of electricity), the following very important question arises: Given the height of the water column above

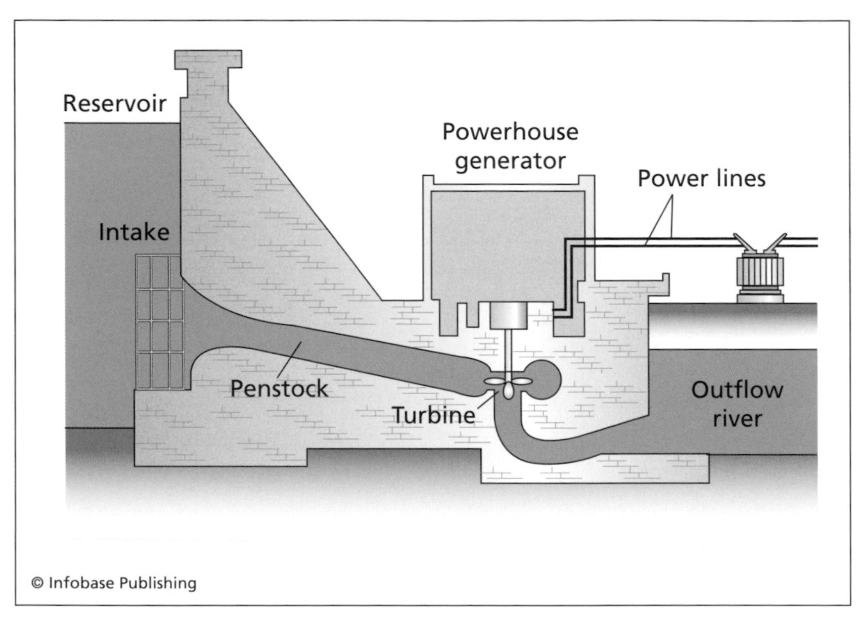

Cutaway view of a hydroelectric plant

the turbine—that is, given the hydraulic head—and the amount of water flowing past the turbine per unit time, what is the most electrical power that a hydroelectric facility can possibly generate? It is an important question, because the answer places limitations on what can be obtained, even in theory, from a hydroelectric station.

When discussing the maximum amount of electrical power that can be obtained from a stream of flowing water, it is helpful to think of the power of the flowing water as the input to the hydroelectric station and the electrical power as the output. The plant converts some of the input into output, but some input is lost to friction and other inefficiencies. Or to put it still another way: The power of flowing water is the raw material on which the turbine-generator system acts; the electricity is the finished product. And because the hydroelectric station only converts one form of energy into another, the amount of power produced by a generating station cannot exceed the amount of power supplied. (This is a consequence of the statement that energy cannot be created.)

For a hydroelectric station, the power of flowing water is most easily expressed in terms of an algebraic equation:

$$P = qha \qquad \textbf{(2.1)}$$

where the letter P represents power, q is the volume of water flowing through the facility per unit time, h is the hydraulic head, and a is a constant, the value of which depends on the units in which P, q, and h are expressed. If P is expressed in watts, q is expressed in cubic meters per second, and h is expressed in meters, then the scaling factor a is approximately 9,800 newtons per meter cubed. (Electrical output is usually not expressed in terms of horsepower, but to find the output of a turbine in horsepower, multiply its output expressed in *kilowatts* by 1.34.)

As one might expect, the actual amount of power produced by a hydroelectric station is always less than the maximum possible amount, but it is proportional to P. If P_{act} is the actual power output of a hydroelectric power station, then

$$P_{act} = eqha$$

where e is the efficiency of the conversion process. (The exact value of e depends on the design of the station, the way that the equipment is maintained, and so forth. In all cases, e is a number between 0 and 1, and the closer e is to 1 the more efficient the station is.) Because P and P_{act} are proportional, and because the formula for P is a little simpler than the formula for P_{act}, the following discussion will use equation (2.1), the equation for P, but the discussion can also be repeated word-for-word in discussing the equation for P_{act}.

Equation (2.1) reveals some very important facts about hydroelectric power. To generate a given amount of electrical power one can use a small amount of water provided the hydraulic head is high enough. Conversely, one can generate the same amount of power with a low hydraulic head provided a large enough volume of water flows through the turbine. Equation (2.1) even shows how the two quantities must vary in order that the power remains constant: h must vary inversely with q. It is not, therefore, the hydraulic head or the volume of water that matters; it is their product.

Equation (2.1) explains why a tall dam with, say, twice the hydraulic head of a shorter dam, will generate twice the power of the shorter dam when operating at the same volumetric flow rate. Taller dams make better use of the available water. Of course, one can also release more water through a shorter dam to achieve the same power output, but there are physical limits on the amount of water that can be released. One cannot, for example, release water through a dam at a rate that is higher than the rate at which the water is replenished from upstream—at least not for very long. On the other hand, taller dams tend to cost much more to build than shorter ones. There is, therefore, a balance to be struck between the initial construction costs and a station's expected power output. (Environmental concerns, which are described in chapter 3, further limit the rates at which water can be released.)

The accompanying diagram (on page 23) is a graph of a curve called an *isoquant,* a line of constant power. The horizontal axis represents the height of the hydraulic head, and the vertical axis represents the volumetric flow rate so the first coordinate of a point on the isoquant represents the value of h, and the second coordinate represents q. (The value a appearing in equation (2.1) is just a scaling factor. It can be ignored—or set equal to 1—for purposes of this discussion. The important characteristics of the curve do not depend on the value of a.) The product of the coordinates of any point on a given isoquant equals the product of the coordinates of any other point on the same isoquant. Each isoquant is, therefore, the set of all values of h and q such that their product equals a given power output. The isoquant illustrates the fact that to produce a given amount of power with a low hydraulic head, one

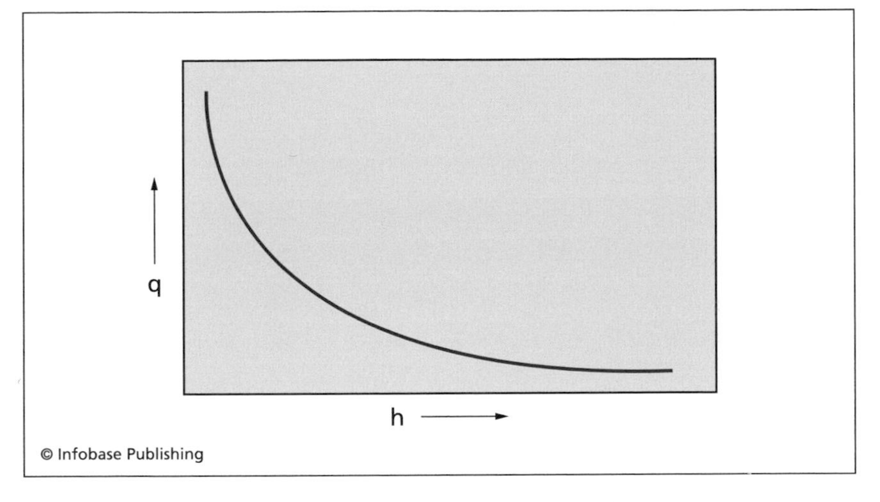

Isoquant curve, a curve representing the set of conditions—hydraulic head and flow rate—that together yield a given power output

must release a very large volume of water. To put it another way, the isoquant is a graphical demonstration that, all other things being equal, taller dams are simply more efficient than shorter ones because they generate more power per unit of water. These are facts of nature. No technology can overcome them. Equation (2.1) provides a framework for evaluating the potential of a particular hydroelectric power project.

TURBINE DESIGN

There was a time during the second half of the 19th century and the early years of the 20th century when many would-be inventors devoted considerable effort to designing water turbines. This was an era when the principles of water turbine design were not entirely understood; there was plenty of room for additional innovation, and large sums of money were being spent developing new water-power sites. Many designs were proposed, and a few found favor with those responsible for designing power stations. In retrospect, most turbine designs fall into one of two categories: Some work best

The remains of a Fourneyron turbine. These turbines played an important part in the industrialization of the United States. *(French River Land Company)*

when operating under high hydraulic heads, and some work best when operating under low hydraulic heads. The higher head turbines are called *impulse turbines,* and the lower head turbines are called *reaction turbines.*

Impulse turbines depend largely on the percussive force of water flowing under a high hydraulic head. In hydropower jargon, a high hydraulic head is usually taken to be in excess of 1,000 feet (300 m). Impulse turbines capture the water with what designers call buckets. A specially designed nozzle directs the water, which shoots out under very high pressure, at the turbine's buckets. The jet of water strikes the buckets and spins the turbine. The basic design for the first successful impulse turbine was patented in 1889 by the American engineer and inventor Lester Allen Pelton (1829–1908). It is called the Pelton turbine.

The story is that sometime during the 1870s Pelton was watching a water turbine spin as water struck the buckets that were placed along its rim. These "buckets" would have been rounded and shallow and looked more like cereal bowls. They were aligned so that the stream of high-velocity water driving the turbine struck each bucket in the middle. The force of impact produced a lot of splashing. As Pelton watched, the turbine became misaligned, and the water began striking the edge of each bucket. The splashing stopped, and instead the water was directed around the curve of each bucket to shoot off the other side of the bucket in a coherent stream. The buckets had reversed the direction of flow of the water. The effect was an increase in turbine speed. Pelton turbines make use of this observation.

The Pelton turbine has been somewhat modified over the years and today has an efficiency of slightly more than 90 percent—that is, it converts a little more than 90 percent of the power in a stream of water. These turbines look somewhat like steel waterwheels. The buckets have a cross section that looks like a rounded "W," and the water is directed toward the wedge in the middle. The wedge splits the stream of water, and the curved surfaces direct the stream backward. As the water emerges from the sides of the bucket it propels the bucket and spins the turbine. This kind of design depends on a high velocity stream of water, a fact that explains why Pelton turbines are usually reserved for facilities with high hydraulic heads.

Sometimes engineers cannot generate a high hydraulic head with which to work. Designers of hydroelectric facilities are constrained by the topography of the land. There may, for example, be a large river flowing through a flat landscape. In this case, it is pointless to construct a tall dam, because it will not constrain the water upstream of the dam. Blocking the flow of the river would only cause the water to flow around the dam. If the river flows through a shallow valley, then a tall dam would fill the valley, possibly putting a large area of land underwater, but it could not create a deep reservoir behind the dam since the valley will constrain the water only until the level of the water reaches the lowest point along the valley wall. Consequently, it is sometimes necessary to also use turbines that are designed to operate under lower hydraulic heads than those characteristic of Pelton turbines. These low- and medium-head turbines are called reaction turbines.

There are several very different designs for reaction turbines. All of them are placed in chambers that are completely filled with water. As the water flows through the chamber it creates a pressure difference; the pressure is much higher on the upstream side of each blade than on the downstream side, causing them to spin. The earliest reaction turbine was created by the French engineer Benoît Fourneyron (1802–67) and is called a Fourneyron turbine. The Fourneyron turbine is mounted on a vertical axis. As the water flows down through the chamber it encounters a set of curved blades that do not spin. These fixed blades direct the water outward against a runner—a round wheel—upon which a second set of blades are mounted. The runner blades are curved in the direction opposite that of the fixed blades. The force exerted by the water on the second set of blades causes the runner to spin. Fourneyron turbines were an important step forward. Fourneyron himself created turbines that operated at 80 percent efficiency, a tremendous technical breakthrough for the time. His most famous turbine weighed only 40 pounds (13 kg), had a runner that was only one foot (30 cm) across, and produced 60 horsepower (45 kW). These turbines were used to power a great deal of industry in Europe, and they also found favor in the textile mills of the eastern United States. In 1895, Fourneyron turbines were installed at the facility at Niagara Falls.

Not surprisingly, engineers have developed a number of other reaction turbine designs since the first half of the 19th century when Fourneyron created the first such turbine. They differ in the

Glen Canyon power plant generators *(U.S. Department of the Interior, Bureau of Reclamation)*

way in which the water flows through the turbine. Fourneyron's design directs the water away from the axis of symmetry of the turbine, while other designs, most notably the Francis turbine, direct the water toward the axis of symmetry, and still other designs are a combination of the two. Designs changed as engineers kept searching for better solutions to perceived shortcomings of earlier designs. (There are even special applications such as the pumped storage facilities, which are described in the sidebar accompanying this chapter, where a Francis turbine rather than an impulse turbine is operated under a high hydraulic head because the Francis turbine can be operated in reverse as a pump, an application for which the impulse turbine is ill-suited.)

In North America by the 1930s, most sites for large-scale hydroelectric development were identified and many were under development. In addition, numerous small-scale sites had also been developed. At the time, these smaller sites offered an economical method of generating mechanical or electrical power.

From 1945 until the early 1970s, many of the smaller sites were abandoned because fossil fuels were so cheap that small-scale hydroelectric power was no longer competitive. In New England, for example, the ruins of small-scale hydroelectric and hydromechanical facilities can still be found along many streams and small rivers. After the energy crisis of the early 1970s, interest in small-scale hydropower facilities began to increase, and now a modest number of sites are developed each year. But because these sites have neither a large hydraulic head nor a large volumetric flow rate, their total contribution would be relatively small even if they were all developed. In some areas, however, they make a welcome contribution to the local economy.

THE WATER SUPPLY

Although most large hydroelectric facilities depend upon dams, most dams are not constructed for the purpose of generating electricity. Often dams are constructed to create a reservoir of drinking water, or as a source of water for irrigation, or for flood control, or for recreation, or to maintain the navigability of waterways, or a combination of these reasons. Dams used at hydroelectric power stations constitute only a fraction of the total number of dams. And not every hydroelectric facility depends on a dam. Niagara Falls is an example of a large-scale hydroelectric station that operates without a dam. Nevertheless, dams are extremely important components of most large-scale hydroelectric facilities. And bigger dams are better than smaller ones, because with respect to power generation the more water that can be stored behind the dam, the more reliably and the more efficiently the station will be able to generate power.

Storing a large supply of water increases the reliability of a hydroelectric facility because a large store of water acts as a buffer in the face of fluctuations in supply. Depending on the location of a particular dam, the rate at which water flows into the reservoir created by the dam may undergo regular seasonal fluctuations, or it may vary sporadically and unpredictably due, for example, to drought. The larger the quantity of water behind the dam, the longer the hydroelectric facility can operate in the face of a shortage in the rate of supply.

Unusually high flow rates can also interfere with power generation at hydroelectric facilities. Floodwaters often transport large amounts of debris, which may require the station's operators to close the *penstocks,* those tubes that connect the turbines with the water upstream of the dam, in order to protect the turbines from damage due to debris. A large volume of water in the reservoir serves to dampen the effects of short-term surges just as it dampens the effects of short-term deficits.

Efficiency in the use of the water resource is enhanced provided the dam creates a reservoir with a high hydraulic head. As previously mentioned in the discussion of equation (2.1), the higher the hydraulic head, the more power can be extracted from each unit of water flowing through the penstocks. But even when it is not possible to construct a facility with a high hydraulic head, it may still be possible, depending on the topography of the site where the dam is located, to impound a large quantity of water. A large water supply will still work to the operators' advantage even without a high hydraulic head, because a large supply of water enables the operators to release large volumes of water without immediately exhausting the supply. Again, as equation (2.1) shows, by doubling the volume of water flowing through the turbine per unit time, the power output can double as well. To be sure, releasing large volumes of water will cause the water level behind the dam to fall—that is, it will cause the hydraulic head to decrease—and a reduction in the

(continued on page 32)

Pumped Storage

Pumped storage technology illustrates all of the ideas described so far in this chapter. As will be seen, it is a technology that consumes more energy than it generates, but it has proven very useful in enabling power producers to meet demand.

To build a *pumped storage facility,* a power producer generally requires something along the lines of a small mountain. A reservoir is built at the top of the mountain, and a second reservoir is built at the base of the mountain. They are connected by penstocks, conduits that allow the water in the upper reservoir to empty into the lower reservoir. Near the base of each penstock, a turbine is installed to harness the power of the falling water. These are specially designed turbines that can operate in reverse, and when they are operated in reverse they function as pumps. As the water flows down from the upper reservoir to the lower one, some of the energy of falling water is converted into electricity. Later, the turbines are used to pump the water from the lower reservoir back up the mountain into the upper reservoir. More energy is always needed to pump the water back to the upper reservoir than is recovered by allowing the water to flow down to the lower reservoir—that is, pumped storage facilities always consume more energy than they produce—so, one might ask, what is the point of building this kind of power consuming machine?

Because electricity cannot be stored in large quantities for later use, it must be produced upon demand. Consequently, timing matters. Pumped storage facilities are valuable because they enable the producer to shift power production, to times when demand is high from times when demand is low. Here is how it works: At night, utilities generally depend upon coal plants and nuclear plants to meet demand because coal and nuclear plants produce very large amounts of power and work most efficiently when operated at a steady output for prolonged periods of time. In fact, at night these plants can easily produce surplus power—that is, they can produce more electricity than there is demand for it. This surplus

power can then be used to "prime" the pumped storage facility. Surplus electricity powers the turbines, which are used to raise the water from the lower reservoir to the upper one. The water remains in the upper reservoir until it is needed. During the day, when demand is high and the coal and nuclear plants cannot meet the entire demand for electricity, the water is released, and the turbines convert some of the energy of the flowing water back into electricity.

Pumped storage facilities are profitable because they enable the power producer to manufacture power when demand and price are highest. In a market with widely fluctuating levels of demand and price, it is often more important to produce electricity at the right time than it is to be a net producer of electrical power.

Tennessee Valley Authority pump storage facility, Raccoon Mountain *(TVA)*

(continued from page 29)
hydraulic head reduces the amount of power obtained from each unit of water. But this reduction will happen slowly, provided the amount of water impounded behind the dam is large enough. As demand lessens, the volume of water flowing through the turbines can be reduced. The water supply will replenish itself, and the process can be repeated during the next demand cycle.

Finally, storing water is, in a sense, equivalent to storing electricity, because water can be converted into electricity by opening the penstocks. Unlike water, electricity cannot be stored in large quantities. The inability to store electricity places a fundamental constraint on the way that power must be produced: Electrical power must be produced simultaneously with demand. The ability to adjust the rate at which power is produced (by opening and closing the penstock gates) is one of the most important advantages that hydroelectric facilities have over other forms of power production. Simple adjustments to the flow rate enable hydroelectric facility operators to respond to changes in demand almost immediately. Of all the methods for generating electrical power, hydroelectric facilities are the most flexible in this regard.

Although engineers became interested in building huge dams for purposes of power production only a century or so ago, attempts to build large dams began much further back in history. The essential challenge of designing a large dam is that large volumes of water exert enormous forces. For a long time, there was no engineering theory that could guide those charged with building these huge structures. Despite this, some early dams display some of the same ideas that one finds in modern dams. The Ma′rib Dam, located in present day Yemen, was built more than 2,000 years ago, and at both ends of the dam the designers included spillways, architectural features that enabled excess water to flow away from the dam. (Water that flows over the top tends to erode—and so weaken—the base of the dam.) Spillways are features of all large modern dams.

And early 14th-century Persians built the Keber Dam, an early arch dam. Arch dams are curved away from the upstream water so that the pressure of the water against the dam is transferred along the dam to its ends. By using an arch the need for enormous quantities of material to hold back the upstream water supply is greatly reduced without sacrificing strength.

Civil engineering, that branch of engineering of most relevance to the construction of dams, grew in large measure out of the work of British engineer and scientist William John Macquorn Rankine (1820–72), whose pioneering studies into the mechanics of soils and other materials formed the basis of a deeper understanding of how dams are able to withstand the tremendous forces exerted on them. Today, engineers have a great deal of insight into the construction and operation of several different types of dams, and as a consequence it has been possible to build ever larger dams with ever more exacting levels of performance.

But as dam-building technology has evolved, the number of undeveloped sites suitable for large-scale hydroelectric power stations has continued to decrease. While hydroelectric power plants will retain an important role in the energy sector of many countries for the foreseeable future, the amount of electricity generated by hydroelectric plants is not expected to increase very much because most of the sites with the potential for generating large amounts of hydroelectric power have already been developed.

BASE LOAD VERSUS PEAK LOAD

Electricity demand fluctuates, and the demand for electricity at any given hour of the day is only partly predictable. Power producers know, for example, that on hot days people will turn on their air conditioners—air conditioning is a very energy-intensive technology—and on those days demand will surge. What they do not know far in advance is which days will be hot. In addition to unpredictable fluctuations in electricity demand, there are many fluctuations

Hoover Dam control room. A small number of people controls one of the nation's great hydroelectric power plants. *(U.S. Department of the Interior, Bureau of Reclamation)*

in demand that are highly predictable. Demand goes up during the day and down at night; during the day demand is higher at 3:00 P.M. than it is at 9:00 A.M.; demand is higher during the day from Monday to Friday than it is during the day on Saturday and Sunday; demand is higher during the summer than it is during the spring, and averaged over several years the total demand for electricity rises steadily. Enormous data sets have been collected for every major U.S. electricity market that make these statements precise.

Underlying these fluctuations in demand is a minimum and highly predictable power requirement that must always be met. Hospitals, for example, consume electricity continuously as do some manufacturing concerns. There are numerous other examples. The electricity required to meet the sum of these minimum power demands is called the *base load*. Base load power plants are those plants that furnish this electricity. Power plants are not distinguished by the type of electricity they produce—the characteristics of the electricity supplied to the grid are the same for all types of power producers—rather, base load power plants are those that furnish an essentially steady flow of electrical power for prolonged periods of time. As a general rule, base load power plants shut down only for maintenance and repairs.

In the United States hydroelectric power plants once provided a large fraction of the nation's base load power. Some hydroelectric power plants are still used to provide base load, but today most base load power is provided by coal and nuclear plants. Coal and nuclear plants are large power producers and work best when operated at relatively steady power levels for long periods of time. By contrast, hydroelectric stations are more flexible and can be operated so as to respond to fluctuating levels of demand. Electricity produced to meet demand fluctuation above the base load requirement is called *peak load* power. (Production is sometimes further divided into intermediate load and peak load, but in this presentation as in many others, it is sufficient to call all power in excess of the base load demand peak power.)

Because electricity cannot be stored, electricity production must vary simultaneously with demand. Matching supply to demand is a difficult technical problem. In a modern market, demand is measured every few seconds, and as it fluctuates, supply must be adjusted accordingly. Matching supply with demand is further complicated by the fact that particular energy technologies may not be available when they are needed. Solar and wind operate independently of demand. The Sun shines and the wind blows (or not) without regard to the requirements of electricity markets. Consequently, they cannot be counted upon to meet peak demand—at least not in the same way that natural gas plants can be relied upon to produce power on demand. If solar and wind producers are producing power when it is needed, then that power can be utilized, but they are unreliable suppliers. Gas-fired and oil-fired power plants are more reliable peak load power producers, but their fuels are very expensive, and they take time to start. Starting a fossil fuel plant is not like flicking a switch. Before they can be brought online they must reach operating temperature. This takes hours. But running them simply because they might be needed is costly both to the ratepayer and to the environment. By contrast hydroelectric power is peculiarly well-suited for meeting fluctuations in demand. As long as water is behind the dam, they are ready to go. They can ramp up or ramp down in a few seconds, because plant operators need only open or close a gate to increase or decrease power production. So even though the technology in use at most hydroelectric plants has not changed in any essential way in many years, the way that many of these plants are operated has changed. (Federal environmental legislation has further affected how hydroelectric plants are operated. See the section "More about Environmental Costs" in chapter 3.)

Costs and Policies

All electric generating stations, whether powered by fossil fuels, nuclear reactions, water, wind, the Sun, geothermal energy, fuel cells, or any other technology, are energy conversion devices. Each technology begins with one type of energy—it might, for example, be kinetic or thermal—and converts it into electricity. Each type of conversion technology has its own environmental and economic costs. When it comes to generating electricity, societies have always been willing to pay all of the costs involved because access to large amounts of reasonably priced electricity make modern life possible. Today the environmental costs of meeting the worldwide demand for electricity are becoming clearer, and the search for alternative technologies is becoming increasingly urgent as more nations adopt westernized energy-intensive lifestyles and begin emitting the enormous amounts of pollution that historically have made Western

Hoover Dam with flag *(U.S. Department of the Interior, Bureau of Reclamation)*

lifestyles possible. There is no easy way to meet the world's appetite for electricity.

To compare different energy conversion technologies in a meaningful way, it is not enough to understand how each technology is used to generate electricity. It is just as important to understand the costs associated with each technology. The purpose of this chapter is to convey something of the costs associated with hydroelectric power.

THE COSTS OF HYDROPOWER

Economists sometimes make a distinction between the words *price* and *cost*. The price of a commodity is what the consumer pays. The cost is interpreted more broadly. By way of example, for a consumer

the price of electricity, no matter how it is generated, is stated clearly on the monthly electric bill that each consumer receives. The cost of electricity includes the cost of building the generating station, which may or may not have been heavily subsidized; it includes damage to the environment caused by generating the electricity; it includes research and development costs; and it includes numerous other items that for one reason or another may or may not affect the price of the electricity. The distinction between price and cost is particularly helpful in assessing the value of hydropower.

Large hydroelectric projects cost a lot. The exact nature of the costs associated with any particular hydroelectric power project depends upon the location of the project, its size, and how it is operated, but the economic, social, and environmental costs of the big projects are often so large that only governments are capable of undertaking them. And because many large hydroelectric projects are paid for and operated by governments, they are often operated in ways that are very different from the ways that commercial generating stations are operated.

Hydroelectric generating stations, their dams, and associated reservoirs are some of the largest engineering projects in the world. Consider the Grand Coulee Dam, which is located in the state of Washington. It is 550 feet (168 m) tall and approximately one mile (1.6 km) long. Although it began operation more than 70 years ago, it remains the largest concrete structure in the United States. To construct the dam required thousands of workers. In fact, when work began in 1934, the consortium of companies hired to build the dam found it necessary to build an entire town, called Mason City, to house some of its employees. While employment levels fluctuated during the construction period, 6,000 workers were employed during 1937. The project did not begin producing power until 1941. The hydroelectric station, which has been upgraded more than once, currently has a capacity of 6,480 MW (*megawatts*) (6.48 gigawatts).

This pattern of long construction times coupled with enormous outlays of money prior to generating any income is still typical of these types of projects. Consider the Nathpa Jhakri hydroelectric power project, one of the largest hydroelectric projects in recent times. It is located on the Sutlej River in India's northern state of Himachal Pradesh. Construction began in 1989, and it began generating power in 2003. Approximately 7,000 workers were engaged in this project at its peak. The Nathpa Jhakri plant has a capacity of 1,500 MW.

Most large hydroelectric plants also require large amounts of land. The Depression-era Hoover Dam, located about 39 miles (62 km) south of Las Vegas, Nevada, created Lake Meade, which covers approximately 247 square miles. The Egyptian Aswan High Dam, which took approximately a decade to build—it was completed in 1970—created a roughly 300-mile (480-km) long reservoir behind it. And the Itaipú hydroelectric power plant, located on the border of Brazil and Paraguay, required the builders to change the course of the Paraná River, the seventh largest river in the world. Only governments can build on such an enormous scale.

To construct the large upstream reservoirs characteristic of these projects requires the power of the government for another reason: Many large hydroelectric projects involve the forced displacement of large populations. In the United States, for the most part, this has not been a characteristic of large-scale hydroelectric development because the largest projects were built in the western part of the nation at a time when the area was still sparsely populated. To be sure, there has been some resettlement, but it is estimated that throughout the history of the United States only about 30,000 people have been forced to relocate as a consequence of hydroelectric plant construction, and the majority of these (approximately 18,000) were displaced as a result of a single project, the Norris Dam, located on the Clinch River near present-day Norris, Tennessee. (The Norris Dam, built during the years 1933 to 1936, forms Lake Norris, with a surface area of about 53 square miles [137 km^2]. The project has a capacity of about 131 MW.) By contrast, the Three Gorges Dam, which is built across the Yangtze River (Chang Jiang) in Hubei Province in China, involved the displacement of more than 1 million people. (The construction of the Three Gorges Dam began in 1993, and it is expected to be fully operational in 2011. It will have a capacity of 22,500 MW.) Nor is the Three Gorges Dam the only such project to cause large-scale dislocations. The Sanmenxia Dam, also in China, involved the relocation of 410,000 people; China's Xinanjian Dam required the resettlement of 306,000 people, and India's Bargi Dam required the resettlement of 113,000 people. To effect such large-scale displacements requires the power of the state.

Large-scale hydroelectric power projects also involve changes to both the upstream and downstream environments. Upstream of the dam a large lake is often formed that changes the ecology all along that section of the river. This degrades the environment for some species and improves it for others. Downstream of the dam, the effects of the hydroelectric plant are more complex and depend on how the dam is operated. Operation of the dam is determined by the interaction of government regulations and market pressures. If, however, the goal is to maintain the environment along the river so that it remains as unaltered as possible, the construction of a hydroelectric power plant presents a serious challenge. The situation is complex and is discussed in more detail in the next section of this chapter.

The effects of the hydroelectric plant on the broader environment are also complex and are due, in part, to the way that hydroelectric power is priced as well as the way that the water behind the dam is utilized. Governments often choose to subsidize the cost of hydropower—that is, the electricity is sold for less than it costs to produce. In one sense this makes the hydroelectric plant a money-loser. But electricity is more than a simple commodity. Inexpensive electricity is a necessary factor for all types of economic growth. In other words, even if a power project never makes money through the sale of electricity, it can still serve an important and positive economic function. The history of the Grand Coulee project clearly illustrates the widespread economic effect of hydroelectric power.

Prior to the time that the Grand Coulee hydroelectric plant began limited operations in 1940, there was no aluminum industry in the Pacific Northwest. (The aluminum industry depends upon large amounts of inexpensive electricity, and aluminum plants will often locate near a source of cheap electricity.) In 1946, only four years after Grand Coulee began full-power operation, 36 percent of the aluminum producing capacity in the United States was located in the Northwest. Not all of the electricity upon which the aluminum plants depended came from the Grand Coulee Dam, but much of it did, and the rest came from other hydroelectric stations that were completed in the same general area at roughly the same time. The Grand Coulee Dam increased the size of the regional economy; it increased the tax base; it drew workers to the area; but it generated no immediate profit from the sale of electricity. In the United States and in some other countries as well, hydroelectric power plants were operated so as to cause economic growth.

The effects of the Grand Coulee Dam on the development of the Northwest were even more widespread than indicated in the preceding paragraph, because Grand Coulee was also built to provide sufficient water to irrigate more than 1 million acres (405,000 hectares) of land. Prior to Grand Coulee's construction there were few farmers in the area. The soil was rich but too dry to farm profitably. After completion of the dam, a large and vibrant agricultural sector was established in the region. Again, the result of the Grand Coulee Dam project was a further influx of workers—this time agricultural workers—and the environment was further changed as large tracts of previously undeveloped land were converted into farmland.

Although each large-scale hydroelectric facility is unique, it is fair to say that the effects of many other large-scale hydropower projects have been somewhat similar to those at Grand Coulee.

One should not conclude from the previous paragraphs that economic development is wrong or unhealthy. Many would consider all of these economic effects to be positive—that is, the economic effects (and the accompanying environmental changes) that resulted from the construction and operation of the Grand Coulee project should be counted as benefits rather than costs. Certainly there are many people whose lives have been enhanced by what the Grand Coulee and other regional hydropower projects have made possible. Whatever one's opinions, the Grand Coulee dam project illustrates that the effects of large-scale hydroelectric projects on the environment extend well beyond the immediate vicinity of the dam and the river on which it depends.

With such high economic, social, and environmental costs, one might question the value of hydroelectric projects, but it is just as important to remember the alternatives. Whatever the costs of hydroelectric power, they are, at least, local—that is, the costs of hydroelectric power development are restricted to the area in which the plant is located. This area might be fairly large—as noted previously, the construction of the Grand Coulee Dam affected the development of the entire Pacific Northwest—but the costs of dam construction and operation are, at least, not global. Compare the costs of hydroelectric plants with those of fossil fuel plants, especially the costs of emissions from fossil fuel generating stations. Today, most electricity is generated by burning coal, and coal technology also has important local costs. The mining of coal involves significant environmental disruption, and mining is a dangerous business; miners are injured and killed on a regular basis in the performance of their jobs. The routine operation of coal plants releases a number of dangerous materials into the environment. Mercury

(continued on page 46)

Global Warming

Global warming, the gradual increase in the average temperature of Earth's oceans and atmosphere, continues to attract a great deal of attention in the popular press and in scientific journals. Global warming is due to changes in the chemical composition of Earth's atmosphere. To see the profound impact of Earth's atmosphere on surface temperature compare Earth with its neighbor, the Moon. When the Sun is shining directly on a region located at the Moon's equator, the surface temperature is about 230°F (111°C). But in the dark, the Moon's surface temperature plunges to -450°F (-233°C). There is no place on Earth's surface with temperatures that approach either of these extremes even though the Earth and Moon are located at approximately the same distance from the Sun. Earth's atmosphere is what accounts for the difference. The density and chemical composition of Earth's atmosphere affects both the amount of the Sun's energy that reaches Earth's surface and the amount of that energy that Earth retains. Change the chemical composition of the atmosphere and the temperature of Earth's surface changes as well. Gases in Earth's atmosphere that effectively retain energy from the Sun and contribute to a warmer climate are called greenhouse gases.

Human beings are altering the chemical composition of Earth's atmosphere—mostly by burning fossil fuels, a process that releases carbon dioxide into the atmosphere. Carbon dioxide is a potent greenhouse gas. Whatever the beneficial effects of heavy fossil fuel consumption—and there have been many, at least in developed countries—it has become clear that the consumption of fossil fuels at current levels releases so much carbon dioxide into the atmosphere that the thermal properties of Earth's atmosphere have changed as a result. Earth now retains more of the Sun's energy than it did in the period immediately preceding the time in which humans began to consume large amounts of fossil fuels. Many of the long-term effects of this change in atmospheric chemistry are not yet clear. What, for example, will be the effects of increasing oceanic and atmospheric temperatures on the powerful ocean currents, such as the Gulf Stream, that distribute thermal energy

about the planet? The answer is currently unknown. But the question is vitally important because changes in the distribution of heat may cause some regions to cool even as the average global temperature increases. The effects of an increase in average temperature can be complex and difficult to predict.

What is clear—because it has been measured—is that the chemical composition of Earth's atmosphere is changing. Because hydroelectric plants convert kinetic energy rather than chemical energy into

(continues)

An iceberg in Drake's passage, off the coast of Antarctica. Although the atmospheric concentrations of greenhouse gases are easily measured—and they are increasing—the effects of these gases are harder to predict. *(Cathy Webster)*

(continued)

electricity, they produce electricity with no greenhouse gas emissions. (Actually, this statement is only true in temperate climates; in tropical locations the situation is more complicated. See "Methane Emissions and Hydropower," later in this chapter.) While large-scale hydropower projects have their own environmental problems, their value must be assessed relative to the presently available alternatives. Many perceive hydroelectric plants as an environmentally advantageous alternative to fossil fuel consumption.

(continued from page 43)

emitted from coal plants in the United States Midwest, for example, has found its way into the bodies of fish throughout the Northeast, and *greenhouse gases* emitted from these same plants contribute to global climate change. The costs associated with fossil fuel consumption are distributed over a much broader area and have more far-reaching consequences than those associated with hydroelectric power. In contrast to the costs of hydroelectric generating stations, the costs of fossil fuel consumption are shared by everyone.

MORE ABOUT ENVIRONMENTAL COSTS

The operator of every hydroelectric power plant must balance profits, production, and environmental concerns. Sometimes these goals coincide, and sometimes achieving one goal means sacrificing another. Achieving the right balance for a particular power plant is a complex problem. With respect to an investor-owned facility, for example, it may not always be possible to maximize profits and minimize environmental damage. There is room for disagreement

Three Gorges Dam on the Yangtze River *(Chang Jiang) in China (Wikimedia)*

among people of goodwill about whether any particular solution is the "right" solution.

Finding a reasonable balance between competing economic and environmental goals is further complicated by the fact that there is no generally agreed upon way to assign a value to environmental damage. There is not even a generally agreed-upon standard as to precisely what constitutes damage to the environment. How, for example, should one determine the value of emissions-free electricity? What value should be assigned to the preservation of a particular ecosystem? Which changes constitute damage and which do not? Even many ecologists tend to answer these questions using quasi-religious language or quasi-scientific ideas, but these kinds of answers do not lead to rigorous solutions; they do not solve the problem of finding the best way to balance the two competing values of meeting the demand for reasonably priced electricity and minimizing the environmental changes that result from the production of that electricity.

Today, a great deal of attention is given to the difficult problem of optimizing the operation of hydroelectric power plants. Finding an optimal solution is further complicated by the fact that every solution, optimal or not, must conform to the regulatory environment determined by the nation in which the power station is located. In the United States, over the last century, federal, state, and sometimes even local governments have, for a variety of reasons, passed laws and adopted regulations to constrain the way that power producers operate. To appreciate some of the ways that these factors affect one another, it is helpful to know a little about the history of the regulation of power producers in the United States.

Dams are heavily regulated. They use a shared resource, in this case a river, to produce a vital commodity, electricity. Ensuring that this is done in a responsible way has long been an interest of government.

Historically, early electricity markets were local monopolies. One company, the local utility, owned the electricity generating stations in its service area; it owned the high-voltage transmission lines that brought the electricity from the location where it was produced to the local market, and it owned the network of low-voltage power lines that connected the high-voltage network to the businesses and homes that depended upon the electricity. (As described in chapter 1, transformers are used to increase and decrease the voltage, a characteristic of electrical current analogous to the pressure of water flowing through a pipe, so that the electricity can be safely transmitted with minimum losses.)

Some of these utilities were privately owned monopolies; other utilities were owned by a local government. The federal government also owned some large generating stations, especially large hydroelectric stations. The reasons for the monopolistic nature of early electricity markets are primarily historical. Early private and municipal companies had to build their own electricity supply and distribution networks if they were going to sell their electricity because originally there was no distribution infrastructure. Suppliers became distributors in order to bring their product, electricity, to market. Moreover, because there was not enough room on a single street for multiple sets of utility poles to accommodate

multiple electricity suppliers, the first supplier to erect a distribution network for a particular region remained the only supplier for that market.

This monopolistic system, which emerged out of the first efforts of Thomas Edison and other technological pioneers, remained in place until the latter decades of the 20th century. One reason for its longevity is that for most of that time it worked well, providing reasonably priced electricity to almost everyone while ensuring reasonable profits for the power producers. But another reason that no alternatives were proposed during these decades is that no one had imagined an alternative. Utility monopolies were widely perceived as "natural monopolies," in the sense that there seemed to be no practical alternative to this way of doing business.

But monopolies were also viewed with suspicion because monopolies of all sorts have sometimes abused their market position and offered consumers limited services at inflated prices. There was fear that a similar situation would arise in the electricity markets. Electricity is, however, too valuable to the economic well-being of the nation to trust to the good intentions of local monopolies. As a consequence, various regulatory bodies were established to ensure that electric rates were "fair" to both consumers and utilities. During this time, hydroelectric plants were usually operated with little regard for the environment.

Beginning in the 1970s the federal government passed a series of laws to constrain the operation of hydroelectric plants in order to minimize the environmental effects these plants have. Of special interest are the 1973 Endangered Species Act, which introduced constraints on the operation of some hydroelectric plants, and the Electric Consumers Protection Act of 1986, which required federal regulators to balance hydroelectric power production with environmental concerns. The effect of these and other later pieces of legislation was to require operators to operate their generating stations in ways that lessened their environmental impacts.

Another important set of changes began when the electricity markets were restructured. The new markets, which are still evolving, are sometimes described as deregulated, but that is not accurate. Instead, beginning in the 1990s, the government instituted a new set of regulations that attempted to introduce competition between power producers—that is, they sought to break up the old natural monopolies. (Certain parts of the electricity infrastructure, most notably high-voltage transmission lines, are, however, still viewed as natural monopolies, and these remain tightly regulated by the Federal Energy Regulatory Commission [FERC].)

The key to the government's attempt at introducing competition lay in a 1996 order by FERC that required owners of high-voltage transmission lines to grant all power producers "nondiscriminatory" access to the high-voltage lines that serviced an electricity market. The idea was that all power producers would use the system of high-voltage transmission lines as a sort of electric highway. Each power producer would have equal access to the highway and each would attempt to produce electricity at a price that would find a buyer. Open access to the high-voltage network would enable multiple power producers to compete for a limited customer base. If one power producer could supply power at a more competitive price than another, then the lower-price supplier would earn profits at the expense of the higher-price supplier. Innovation, it was hoped, would flourish.

At least that was the theory. In the jargon of the industry, a competitive market provides "price signals" that investors can use to guide their investment decisions. The price of the electricity—not its cost—would guide the way that the market evolved. This new system has had a number of intended and unintended effects on the electricity markets.

One consequence of the restructured markets was that some hydroelectric plants were increasingly used for peak power production. Unlike most other power generating technologies, hydroelectric power works equally well providing peak power or base load power. Base load power is highly predictable and so utilities generally enter into long-term contracts with power producers, mostly coal and nuclear, to provide base load power. Some hydroelectric plants are used to provide base load power, but peak power, which is sold on an hour-by-hour basis, is generally sold at a premium. Provided that one can sell enough peak power, there is more profit selling peak than base load power, a clear price signal that the peak power market is the preferred market.

Four economists, Matthew Kotchen, Michael R. Moore, Frank Lupi, and Edward Rutherford, have published "Environmental Constraints on Hydropower," an interesting study of the effects of regulation on the price and the costs of hydroelectric power production. They sought to quantify what happened when, as a condition of a relicensing agreement between FERC and the Consumers Energy Company, Consumers Energy changed the way that it operated two medium-sized hydroelectric plants on the Manistee River in Michigan. The plants in question are the 20.1-MW generating station at Tippy Dam and the 17-MW generating station at the Hodenpyl Dam. Prior to the relicensing agreement both facilities were being used to produce peak power.

If no thought is given to the environment, the most profitable way to produce hydroelectric power is to leave all the water impounded behind a dam and release it only intermittently. In particular, this means releasing little or no water during the evenings when demand and prices are low, and releasing large amounts during the day, when demand and profits are high, and for a while, Consumers Energy operated its dams in just that way. But that operating regime caused large fluctuations in water levels and flow rates on an almost daily basis. As part of its relicensing agreement, Consumers Energy agreed to operate its dams in so-called run of river mode—which is another way of saying that it would monitor the amount of water flowing into the reservoirs and release it through the dams at roughly the same rate. The reason that FERC wanted run of river operating mode is that releasing large amounts of water during periods of peak demand and withholding it otherwise causes a fair amount of environmental damage downstream. In a single day the water levels downstream of these dams varied from drought level to flood stage and back again. Run of river, on the other hand, creates downstream conditions that are more similar to what would exist if the dams were not on the river.

Methane Emissions and Hydropower

Hydropower is often described as an emissions-free source of electricity and in temperate areas of the world this is certainly true, but the situation is more complicated in tropical regions. Deep in the reservoirs that form behind large tropical dams, bacteria feed on organic matter in low-oxygen environments. One of the by-products of this process is methane, which is released by the bacteria. At the high pressures that exist deep beneath the surface, the methane simply accumulates in the water. But when this water passes through turbines and encounters the low (atmospheric) pressure on the downstream side of the dam, the methane bubbles out of the water in much the same way that carbon dioxide bubbles out of soda when the bottle is first opened. This has important environmental consequences because, once in the atmosphere, methane retains heat far more efficiently than does carbon dioxide—that is, methane is a far more potent greenhouse gas than is carbon dioxide (about 20 times as efficient)—and for some dams, the amount of methane released into the atmosphere is substantial. Scientists at Brazil's National Space Research Institute estimate that for some Brazilian hydroelectric facilities, the methane released in this way contributes more to climate change than would the carbon dioxide released by a fossil fuel plant with a similar power output. In fact, they estimate that each year worldwide, the total effect of the methane released during dam operations in tropical

Under the new agreement, the two facilities produced less electricity during peak periods but began producing energy during off-peak times. The consequences of the new operating procedures were complicated: First, the belief that switching to run of river operating mode would be better for the environment was justified. Chinook salmon runs on the Manistee River more than tripled, from 100,000 per year to approximately 370,000 after the operators switched to run

environments has the same effect as releasing 800 million tons of carbon dioxide, an enormous environmental load.

The solution, which is also being developed at the National Space Research Institute, is to capture the methane before it passes through the dam. An intake pipe will pull the methane-rich waters that exist deep within the reservoir up to a natural gas plant that would be built on-site. (Natural gas is almost pure methane.) As the methane separates from the water, it is captured and burned to produce electricity. The water, now nearly methane-free, is returned to the reservoir. Of course, the carbon dioxide produced at the plant by burning the methane is also a greenhouse gas, but its effect on the climate is far less than that of the methane from which it was derived. Moreover, the mass of the carbon dioxide produced by burning the methane is roughly equal in mass to the carbon dioxide removed from the atmosphere by the plants that produced the organic matter on which the deepwater bacteria fed. Returning the carbon dioxide to the atmosphere simply "completes" the cycle, so in this sense the Brazilian scheme is approximately carbon neutral. In areas of the Amazon where the methane content is exceptionally high, the fossil fuel plants that would be built to capture the methane could increase the power outputs of their respective hydroelectric facilities by as much as 50 percent.

of river. Second, Consumers Energy's costs for operating the dams rose by approximately $300,000 per year. FERC had predicted that the change in operating mode would enable Consumer's Energy to maintain total power output. This, too, proved to be correct, but some of that energy was now produced when there was less demand—and so less profit for the producer—than under the peak-production mode. The difference in timing accounts for what is essentially the $300,000 loss. Most interesting was the impact of changing the operating mode of the dam on Consumers Energy's greenhouse gas emissions.

Consumers Energy's first responsibility is to meet the total minute-by-minute demand for electricity on the part of its customers. Their choices for meeting this demand were hydroelectric power or fossil fuel plants. Having cut back hydroelectric power when demand was high, Consumers had to generate more peak power with fossil fuels. But because Consumers increased hydroelectric power when demand was low, they cut back on that portion of base load power produced by fossil fuel generating stations. The difference, however, was about more than timing. Base load generating capacity is, in this case, largely done by coal-burning plants. Peak load generating capacity is generally met with natural gas plants. Burning coal releases more carbon dioxide per unit of heat produced than does burning natural gas. By changing power-production schedules, Consumers Energy burned less coal and more natural gas. As a consequence, the run of river operating regime also cut, if only a little, air pollution levels and the emission of greenhouse gases.

Was it worth it? The answer depends on the relative values that one places on the environment, company profits, and the price of electricity. With respect to hydroelectric power, each plant is unique and requires its own analysis. Even after 100 years there is still more to learn about the best ways to employ these valuable resources.

THE FUTURE OF HYDROPOWER

Worldwide, the number of remaining sites suitable for traditional large-scale hydropower development is not large and cannot be increased. There are more medium-sized sites with smaller hydraulic heads and lower volumetric flow rates such as the Tippy Dam, described in the previous section, and there are many sites suitable for microhydropower development. Microhydropower systems are defined to be those that generate no more than 100 kW, but they often generate much less. The difficulty with medium and smaller-sized facilities is that it takes so many of them to produce the same amount of power as a single large unit. Looking back at equation (2.1), one can see that to produce a large amount of electricity requires a high hydraulic head and a large flow rate. To see why large facilities are better in this respect, compare a larger and a smaller generating station. Suppose a smaller station has half the hydraulic head and half the volumetric flow rate of a larger facility. The result is that the smaller facility can *at best* produce only one-fourth the amount of power of the larger facility. Small facilities are not just small in size; they are inefficient in the sense that they require more water to produce the same number of kilowatts. Because of these inefficiencies there are hard limits to what can be done with smaller facilities.

In the early days of the United States' Industrial Revolution, engineers quickly identified locations along rivers in the eastern part of the country that combined large vertical drops over short distances with good flow rates. Factories sprouted in these places as fast as the engineers could build the facilities needed to harness the rivers. There was a limited number of such locations, of course, and they were soon developed. These technological pioneers recognized these locations as valuable natural resources to be exploited for the public and the private good, though not necessarily in that order. Today, as the consequences of the fossil fuel economy become increasingly clear, there is renewed interest in developing hydropower as well as heightened interest in the more effective use of hydropower. Both areas are of some importance for economic and environmental reasons, but the value of conventional hydropower is limited. Most countries simply do not have a sufficient number of sites to produce a large percentage of their power from flowing water, and in developed nations most of the best sites have already been developed. In most cases, countries interested in meeting additional demand with reasonably priced emissions-free power will have to look to other power-production technologies.

Electricity from the Oceans

Wave Power

Anyone who has seen pictures of ocean waves crashing against the shore knows that there is a great deal of power in waves. They can reduce cliffs to rubble and destroy large ships. These examples illustrate both the advantage and the disadvantage of wave power: Waves are both powerful and (usually) destructive. A machine designed to convert the energy of waves into electricity could, at least in principle, produce a great deal of electrical energy in such an energetic environment. The main barriers to exploiting this resource are that (1) wave power occurs in surges, and (2) the wave environment is a destructive one. It has so far proven difficult to design a device that can withstand the routine violence of the sea and still produce power reliably. But technology continues to improve. This chapter considers three methods for converting wave energy into electrical energy.

Winter storm swells, North Pacific. The ocean can be a very energetic environment. *(NOAA)*

Before describing these conversion schemes, it is worthwhile to consider the waves themselves. A wave is a disturbance in the surface of the ocean. The types of waves of importance here are generally caused by wind blowing across the ocean's surface. Far from shore, if the wind blows hard enough and over a long enough distance it can create substantial disturbances (waves) that move across the surface in a motion that some describe as rolling. It is the disturbance that moves forward; the water of which it is momentarily composed simply oscillates in a (more or less) up and down motion—that is to say, it is the wave rather than the water that moves across the ocean.

Waves are classified according to three characteristics: (1) the *amplitude* of the wave, which is defined as one half of the vertical distance from the wave's peak to its trough, (2) the wavelength, *L*, of a wave, which is defined to be the distance from one peak to

the next, and (3) the wave period, T, which is the time that elapses between the passing of wave crests. The speed with which a wave travels depends in part on how deep the water is, but in deep water wave speed is a simple function of the wavelength and the period:

$$v = L/T$$

where v represents the speed of the wave.

Wind generates waves of different wavelengths and periods and so, as the preceding equation indicates, ocean waves travel at different speeds. A faster wave may overtake a slower wave, and as it passes through the slower wave they will briefly combine to form a new wave with an amplitude that is greater than the amplitude of either of the individual waves. The faster wave continues to move forward and eventually the two waves separate; their characteristics—amplitude, wavelength, and period—will be the same after the encounter as they were before. This phenomenon is called interference, and it helps to explain why ocean waves may vary in height and distance from one another.

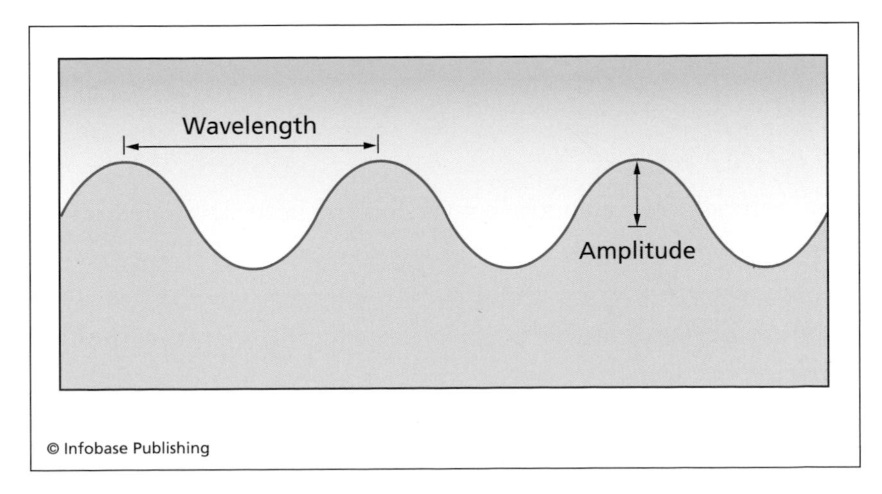

Sine wave showing amplitude and wavelength

Waves, having formed in a windy region of the ocean, last a surprisingly long time. It is not uncommon for waves to roll across the sea for many hundreds of miles even when the wind that created them ceases to blow. As waves travel across the ocean they slowly dissipate. Waves with shorter wavelengths fade first so that a group of waves that have traveled a long way across the ocean will be composed exclusively of waves of longer wavelengths. Near where the waves are formed, however, there will be waves of many different wavelengths, and this makes for a more turbulent sea.

Finally, the energy of the wave—and it is the energy that matters to those attempting to convert wave energy into electricity—is proportional to the square of the amplitude of the wave. The equation that describes this situation is

$$E = pA^2$$

where E represents the energy of the wave, A represents its amplitude, and p is a constant of proportionality whose value depends on the units used to express energy and amplitude. This equation is important because it shows that if one doubles the amplitude of the wave, the amount of energy in the wave increases by a factor of four. Therefore, small changes in the height of waves represent large changes in the energy of the waves.

SEA SNAKES

One method of wave energy conversion has been developed by the company Pelamis Wave Power. Their device is called the Pelamis Wave Energy Converter, and it has already been deployed in small numbers. (Pelamis is the name of a snake that swims across the surface of the sea.)

The Pelamis Wave Energy Converter, or "the Pelamis" for short, which is sometimes called the "snake" because of its shape and motion, consists of four long cylindrical sections. The sections are aligned one after the other and connected by three flexible joints to

Pelamis wave energy farm *(Pelamis Wave Power)*

form a segmented string-like assemblage almost 12 feet (3.5 m) in diameter with a total length of approximately 500 feet (150 m). The Pelamis is designed to float at the ocean's surface while held in place by a tether.

Pelamis wave farms will usually be located a few miles from shore, where the waves are more predictable, and in waters deep enough so that the waves are unaffected by interactions with the seafloor. These near-shore locations are important because the cost of equipment needed to carry the power from a Pelamis wave farm back to shore is high, and each unit must be towed to shore for service and repair. Siting the "wave farm" as close to shore as possible—but not so close that the waves are weakened by interactions with the seafloor—reduces these costs. The first Pelamis wave farm, a small operation located three miles (5 km) off the Portuguese coast, began operation in the latter half of 2008. Another wave farm is planned for Scotland, where Pelamis converters will be used to provide power for the Orkney Islands. Pelamis facilities are modular—that is, to obtain more power one simply adds more converters.

The devices work in the following way: They are tethered to the ocean floor so that they can swing perpendicularly to incoming waves. As waves pass along the length of the device, the buoyancy force, the force exerted by the water against the body of the snake, varies from point to point. The Pelamis, because it is jointed, tends to sag in the troughs of the waves and arch upwards where the crest is passing. The Pelamis's motion is not really snakelike—its body is far too rigid—but the cylinders do bend at the joints. As a joint begins to flex, a great deal of force is exerted at the joint by the massive steel cylinders, and this force is used to power a type of pump called a *hydraulic ram*. The hydraulic ram drives oil through a motor, which performs a function analogous to that of a turbine in a hydroelectric facility; the hydraulic motor drives a generator, which produces the electricity. The power is sent to a cable below the unit. The cable carries the electricity to shore.

In concept the Pelamis wave energy converter is almost, but not quite, that simple. Because the joints are powered by passing waves, they flex only intermittently and irregularly. Without additional modification, the output from the Pelamis would be too unsteady to be useful. The engineers' solution is to employ something called an accumulator. The accumulator, which is placed between the motor and the hydraulic ram, captures the oil and releases it to the motor at a steadier rate. This extra step smoothes the power output of the device.

So far Pelamis has found buyers because the technology shows promise. These devices produce power with zero emissions. They have a low visual profile, which is an important advantage since one of the main objections to wind energy is that the equipment is

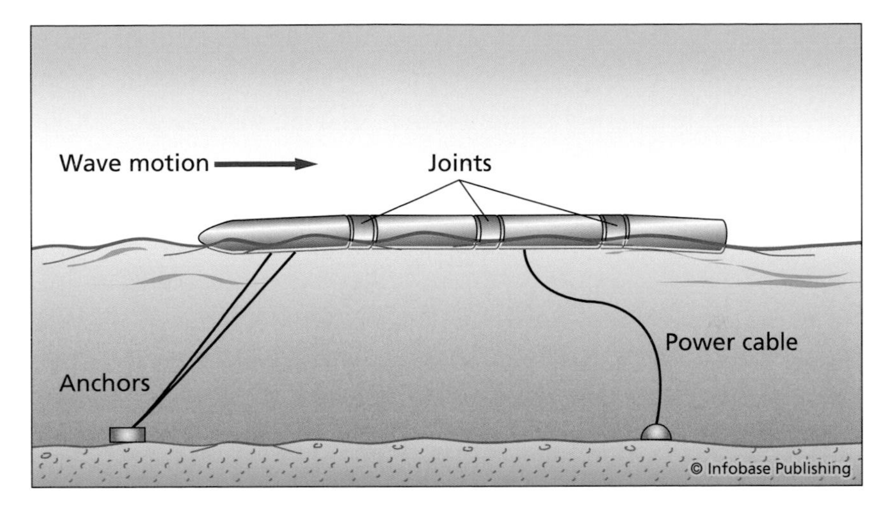

As waves move past the Pelamis, they cause it to flex, and as it flexes it generates electricity.

visually intrusive. In contrast to the tall towers of wind turbines, Pelamis converters float low in the water and are located miles from land. And, as previously mentioned, these devices use a resource that is more predictable than wind. In theory, Pelamis wave farms could produce large amounts of power while occupying a smaller area of Earth's surface than wind farms with similar outputs. Even now, for certain remote applications, the Pelamis system can make an important contribution. Those are some of the main advantages to the Pelamis system.

There are disadvantages to the system as well. At present these wave-energy converters are not economically competitive with more conventional power sources. As of this writing, power from a Pelamis unit costs about twice that of power generated by wind turbines, but Pelamis representatives claim that as production volumes increase costs will quickly decrease. This is surely true, although it is not as clear that costs will decrease enough to make Pelamis power competitive with wind—never mind conventional sources of power, which are often significantly cheaper than wind.

The second disadvantage involves changes in energy availability. Each snake has a rated power output of 750 kW, but unlike more conventional generating stations—that is, coal, natural gas, oil, or nuclear—the amount of power each unit actually produces is variable and in most ways is beyond the control of the operator. Although waves are more predictable than wind, wave energy also varies from day to day and according to the season. Some of this variation is predictable, and some is not. Pelamis technology decouples supply and demand in the sense that a wave farm may produce large amounts of power when demand is low and small amounts of power when demand is high. Once the site is chosen and the units deployed, the level of power production is largely out of the operator's control.

To their credit, Pelamis designers have found ways to increase the efficiency of their devices during periods of low wave activity and so lessen the effect of the inevitable variations in wave energy. But even for the ocean test site used by Pelamis for data acquisition, the projections that they published indicate that a wave farm at their site would generate more than twice as much power during winter months as during summer months. In other words, producing the same amount of power during summer as during winter would require deploying many more units than would be needed to meet wintertime demand. Alternatively, one could deploy other technologies in order to assure adequate power supplies when the wave energy converters fall short. Both strategies are expensive. The Pelamis is also designed to shut down during periods of intense wave activity to prevent damage to the unit. This saves the wave farm, but increases its unreliability as a source of energy.

These advantages and disadvantages need to be evaluated relative to the other options that are available. True, wave energy fluctuates, but it does not exhibit as much day-to-day variation as does wind. Buyers must choose the best technology from the available alternatives. They do not have the luxury of choosing the best technology imaginable.

The Pelamis converter may prove useful to buyers willing to pay the premium necessary to diversify their energy base—especially if their goal is to produce some power from waves, some from wind, and some from solar, for example. By distributing one's risk among several different intermittent energy sources, one lessens the probability that there will be power production shortfalls since it is less likely that the wind, waves, and Sun, for example, will all fail simultaneously. But this approach also increases costs since it requires the power producer to build parallel systems, each using a different energy source and each system built with enough capacity to meet the level of demand that will occur when one or more of the other technologies fails to produce. By distributing risk in this way, the system operator can decrease the probability that it will be unable to meet the demand for power. Alternatively, fossil fuel plants can be used to provide the necessary margin of safety. What is certain is that the system requires backup, and backup systems cost money. It is a truism that money spent on electricity is money diverted from other uses, such as health care, education, transportation, and even energy research. Whether consumers and taxpayers are willing to pay the extra costs involved in deploying the Pelamis and other similar systems remains to be seen.

BLOW HOLES

The oscillating water column (OWC) energy converter is the name of the technology described in this section. The first of these devices to provide electricity to the grid is called the Limpet and is located on the isle of Islay off the coast of Scotland. The Limpet plant began generating commercial amounts of electricity in the year 2000. (Scotland, which invests heavily in renewable energy, is also where Pelamis Wave Power is located.) The Limpet's OWC technology is produced by a company called Wavegen and in concept the idea is elegant and simple.

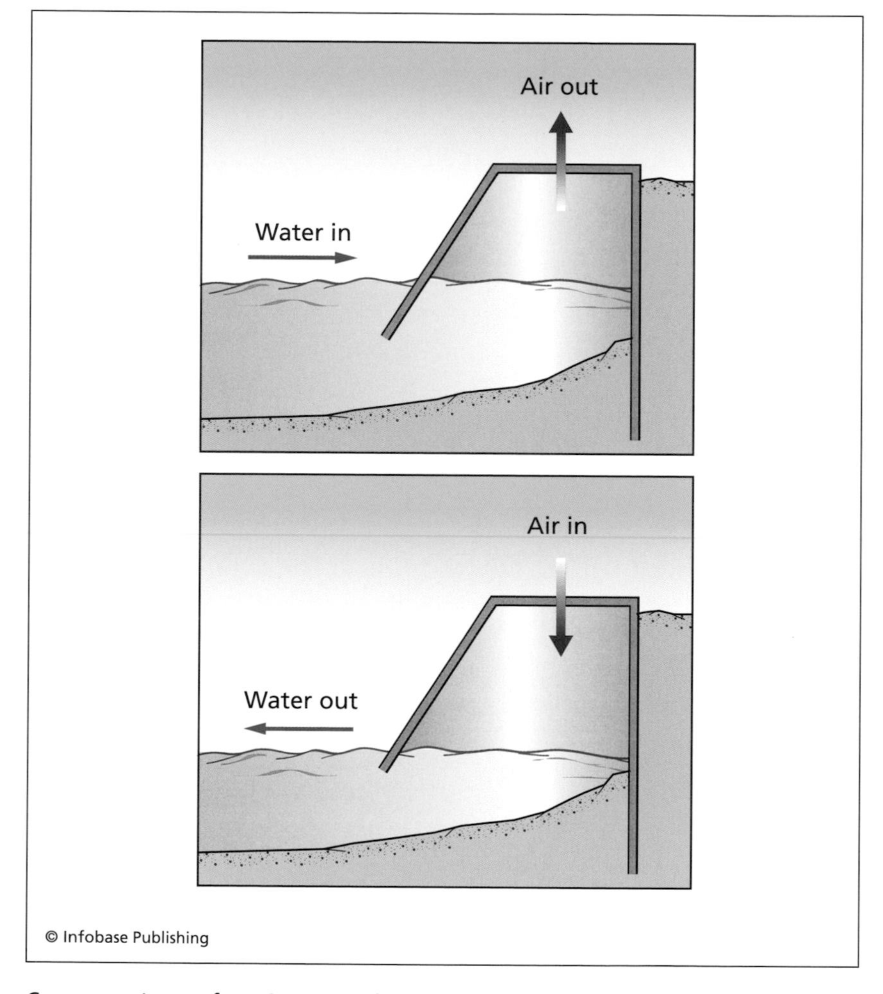

Cutaway views of an OWC. As the wave approaches air is forced out of the opening on top, and airflow reverses as the wave recedes.

To see how an OWC energy converter works, imagine a bicycle pump. Pushing on the handle of the pump drives the piston downward through the cylinder. The piston pushes the air forward and causes it to rush out the other end of the hose, which is attached to the bottom of the cylinder. If the cross-sectional area of the hose were the same as the cross-sectional area of the piston, the air would flow out of the pump only as fast as the pump handle moves forward, but pumps are not designed that way. Instead, the cross-sectional area of the hose is many times smaller than the surface area of the piston. As a consequence the speed of air flowing out of the hose is many times greater than the speed of the piston. There is an easy algebraic equation to describe this situation:

$$S_p A_p = S_h A_h \qquad \textbf{(4.1)}$$

where S_p and A_p represent the speed of the piston and the area of the piston, respectively, and S_h and A_h represent the speed of the air as it passes through the hose and the cross-sectional area of the hose, respectively. This equation was known to Leonardo da Vinci from his studies of water moving through canals of variable cross sections. (From a mathematical point of view equation [4.1] is identical to equation [2.1].)

Equation (4.1) shows that if the cross-sectional area of the hose, A_h, is small then the speed, S_h, must be large in order that the product of the two terms equals $S_p A_p$. In fact, even if S_p, the speed of the piston, is not very fast, the product $S_p A_p$ will be large as long as A_p is large enough. Consequently, given $S_p A_p$, in order to force the air to rush very quickly out of the hose, one need only choose a hose with a small enough cross section. The entire process works the same in reverse. If one pulls the handle of the pump back so that the piston moves up the tube and away from the hose, air will rush up the hose and into the piston at the speed predicted by the preceding equation.

OWC technology works on the same principle as the pump. First, a large structure is built in an area of strong wave action. The structure is boxlike in shape and has two openings. The first is a comparatively small hole at the top of the box. The second "hole" is the bottom of the box, which is left open so that water can flow in and out. The walls of the structure extend down below the waterline so that the ocean forms an airtight seal at the base of the box.

Each incoming wave causes water to flow under the walls and up into the box. The rising water level inside the box displaces the air in the box, which rushes out of the hole (or holes) at the top. (The water acts like the piston in a pump.) As the wave recedes, the water inside the box flows out of the bottom and creates a region of low pressure inside the box. Air flows back down into the box through the smaller opening (or openings) at the top. (This is equivalent to pulling up on the handle of the pump and withdrawing the piston from the tube.) By adjusting the relative sizes of the box and the opening (or openings) at the top, it is possible to create a powerful wind flowing through the top first in one direction and then in the other. The situation is completely analogous to adjusting the air speed through the hose of a bicycle pump by adjusting the relative cross-sectional areas of the piston and the hose.

To convert the up-and-down oscillations of the waves into electrical power, the next step is to place a turbine in each opening at the top of the structure. As air rushes in and out through the blades of the turbine, the turbine converts the linear motion of the air into rotary motion, and the rotary motion of each turbine is used to drive a generator. Technically, a significant problem with this design is that the airflow is continually reversing direction—out of the box when the wave is rushing forward and into the box when the wave is receding. The solution is to use a turbine that turns in the same direction regardless of the direction of airflow. These turbines now exist.

It is important to note that Wavegen is not the only company engaged in developing wave power using an oscillating water column. The Japanese government has long supported research into a similar idea called the Mighty Whale, and an Australian company, Energetech, has further modified the ideas of OWC technology used at the Limpet plant by employing a parabolically shaped wall to focus the energy of a large segment of an oncoming wave into a small area with the goal of increasing the efficiency of the device—that is, increasing the amount of electrical energy generated per wave.

As with other wave energy converters, OWC technology produces no greenhouse gas emissions. The turbine, the main moving part, can be easily removed for repair or maintenance because it is on land. The converters can be placed along the shore, as with Wavegen's Limpet, or near the shore, as with the Mighty Whale and the Energetech device—almost anyplace where there is significant wave action. Because waves are more regular than wind, the output of an OWC design is, in theory, somewhat more predictable than that of a wind turbine.

Most of the drawbacks of OWC converters are similar to those of the Pelamis: Power from the OWC converters is more expensive than that produced by other more conventional power plants, but at least part of this disadvantage would disappear if the units were mass-produced. A second disadvantage is that, as with the Pelamis, the OWC's output is dependent on the level of wave energy, and wave energy varies day by day and according to the season. The amount of energy available for conversion does not, however, depend on the level of demand for electricity. Therefore, the power that these devices produce may not be available *when the power is needed*. Therefore, compared to more conventional power sources such as fossil fuel plants, OWCs are not especially reliable. There is nothing to be done about this. It is a characteristic of OWC and other similar technologies. Finally, output per unit, even under the best of conditions, is relatively small because by the time a wave is near the shore much of its energy has dissipated. Interactions between the wave and seafloor weaken the wave, and OWCs are built along the shore or in shallow water. As a consequence, it would be necessary to construct many OWC plants in order to make a substantive contribution to the power supply of any large energy market. But if its value to the larger economy is limited, an OWC plant can make a real contribution to more isolated island markets, which is, after all, exactly where the first unit, the Wavegen unit, was constructed.

THE ARCHIMEDES WAVE SWING

The last wave energy converter to be considered in this section is called the Archimedes Wave Swing (AWS), and it, too, is being developed in Scotland. The company building the AWS is AWS Ocean Energy Ltd. The AWS operates under somewhat different principles than the Pelamis or the OWC. The AWS is completely submerged so that waves pass over it—recall that waves pass under the Pelamis and crash into the Limpet.

The AWS operates on a pressure difference caused by each passing wave. Pressure beneath the ocean is a simple function of depth: the deeper one goes, the higher the pressure one experiences. The relation between depth and pressure is summarized in the following algebraic equation, where P represents pressure, w represents the weight of the water per unit volume, and h represents the distance to the surface:

$$P = wh$$

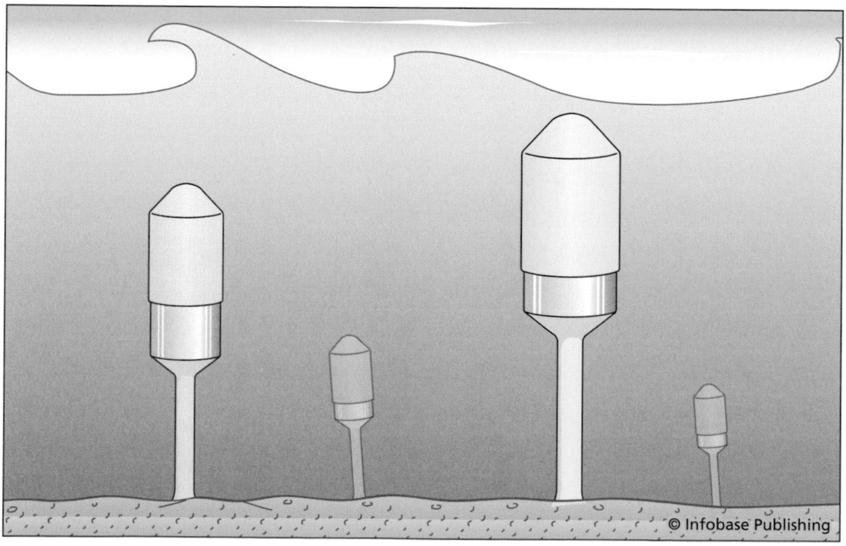

© Infobase Publishing

The Archimedes Wave Swing is driven by the pressure changes of passing waves.

Water does not compress very easily, so at the depths at which the AWS operates, w can be taken to be a constant. Changes in pressure are, therefore, proportional to changes in h, the height of the column of water. Pressure changes caused by passing waves are what activate the AWS. The larger the wave, the larger the change in pressure, and the more electrical energy the AWS generates.

Here is how the AWS works: The designers of the AWS created a large cylinder filled with air. (The support structure for the cylinder is firmly attached to the seafloor.) The cylinder consists of a moveable upper section and a fixed lower section. The air acts as a sort of spring that serves to restore the cylinder after it has been compressed by pressure increases caused by passing waves. When the peak of a passing wave is positioned above the cylinder it creates a momentary increase in h, and the resulting increase in h forces the upper part of the cylinder to descend. The air inside the cylinder compresses and the pressure increases until the pressure inside the cylinder is sufficient to balance the pressure outside the cylinder. The crest of the wave continues moving forward and away from the AWS. The peak, of course, is followed by the trough of the wave. As the trough passes over the AWS, the height of the water column above the AWS takes on its minimum value. Consequently, the pressure on the AWS is also at its minimum value. The air inside the cylinder, which was compressed by the peak of the wave, now expands outward, pushing the upper part of the cylinder upward until the pressure inside the cylinder balances the pressure outside. The cycle is repeated with each passing wave. The AWS is a large piston activated by pressure differences caused by incoming waves. The up-and-down motion of the piston drives a *linear generator,* a device that converts the piston's motion into electricity. As long as the waves continue to roll in, the AWS will continue to generate electricity.

Interest in wave energy is driven by several factors: emissions-free electricity, the large amount of energy in ocean waves, the fact that waves are somewhat more predictable than wind, and what companies call the lower "visual profile" of wave farms relative to wind farms. Large wind turbines are visible from far away because they are so tall. In some places, people object to the visual "intrusion" of wind towers and work hard to prevent the wind turbines from being erected. By contrast, all wave energy converters are less visible than wind turbines, and in the case of the AWS, once deployed, they are impossible to see without diving gear. The drawbacks associated with OWC and Pelamis technology, described previously, also apply to AWS technology: AWS technology is dependent on the supply of waves, a factor beyond the control of the operators, and wave energy may be low when electricity demand is high. And the cost of AWS technology remains an issue just as it does for the Pelamis and OWC technology. Electricity produced by these technologies costs more than wind, and wind costs more than conventional sources. All of which illustrates just how difficult it is to replace—rather than simply supplement—more conventional power sources, especially coal, which in the United States, Germany, and China, three of the world's largest economies, remains the most important source of energy for generating electricity. It is by no means certain that these wave-energy conversion technologies will ever provide more than a tiny fraction of the electricity required by most large industrialized countries. Whether these energy conversion technologies eventually flourish will depend on how competitive they are with respect to all other power technologies.

Tidal Power

Harnessing the tides could, in theory, produce enormous quantities of electricity. So far, however, the electrical energy obtained from the tides has been miniscule. The difference between theory and practice could not be more extreme. This chapter begins with a short description of tides and then examines two schemes for converting tidal energy into electrical energy.

Tides result from the deformation of the Earth's oceans due to the gravitational tug of the Moon and the Sun. The Moon, because it is so much closer to Earth, exerts a gravitational pull on Earth's oceans that is about 2.2 times as strong as that exerted by the more massive and more distant Sun. When the Earth, Sun, and Moon all lie along a straight line, the effects of the Sun and Moon are additive—that is, their gravitational forces combine to amplify the tides. These are called spring tides. When the Earth, Sun, and Moon lie at the corners of a right triangle, the gravitational forces exerted by the Sun and Moon partially cancel one another, and the tides are reduced in magnitude. These are called neap tides.

If Earth were a smooth sphere covered to a uniform depth by one vast ocean, tides would be a simple matter to describe: Bulges would form in the ocean due to the interaction of the ocean with the Sun, the Moon, and the rotation of the Earth. These bulges would remain more or less aligned with the Moon, so as Earth rotated about its axis—causing the Moon to appear to rise and set—the bulges would move across the planet's surface. The tides would be higher when the Earth, Moon, and Sun were aligned and lower when they were not aligned. It would be that simple.

In practice, tides are far more complicated. The main reason is the interaction of the oceans with Earth's topography. The depth of the ocean varies from several miles to a few inches. The seafloor is often rough on a small-scale and on a large scale, and is comprised, in part, of mountains, cliffs, and valleys, all of which serve to channel the oceans' water as it flows steadily in response to the forces acting upon it. In addition to the uneven seafloor, the continents that bound the oceans have complex shapes. Harbors, reefs, rivers, and barrier islands, for example, all have localized effects on the height of the tides. These geographical features can redirect the tidal motions in their vicinity and cause parts of the ocean to flow as if they were large rivers. At the Bay of Fundy in Canada, the rising tide pours 70 billion cubic feet (2 billion m^3) into the bay twice each day before rushing out again. Less dramatic but just as significant, there are other locations in the ocean where tides are virtually absent. Tides are, therefore, a local phenomenon. Each potential site is unique; each site requires its own analysis.

Throughout the years, tidal energy enthusiasts have described the tremendous amounts of energy associated with these large and regular water movements and made extravagant claims regarding the economic potential of harnessing tidal energy. But there is a difference between theory and practice. The obstacles involved in harnessing tidal energy are formidable. Consequently, tidal energy has so far made only a tiny contribution to world energy output. That may soon change, however, as new approaches toward converting the ebb and flow of the tides into electricity are developed.

THE FRENCH AND CANADIAN PROJECTS

The technology discussed in this section, a method that employs dam-like structures called tidal barrages, is hardly new. The Romans built conceptually similar structures that harnessed tidal energy to mill grain. The Roman scheme worked as follows: A dam-like structure was built in a small inlet. The dam was equipped with a gate called a sluice. As the tide rose, the sluice was opened, allowing water to flow through it and accumulate behind the dam. At high tide the sluice was closed, trapping the water behind the dam. As the tide ebbed it created a difference between the height of the water behind the dam and the height of the sea in front of the dam. When the tide was low enough—and the hydraulic head was high enough—the Romans released the water behind the dam in a controlled way. As the water flowed back to the sea it was used to drive a waterwheel. The waterwheel powered a mill that ground grain. Unlike a waterwheel driven by a river or stream, the tidal mill only worked twice each day, because only two tidal cycles occur each day. (A tidal cycle—high tide to low tide and back to high—occurs approximately once every 12 hours and 25 minutes.)

The mills were advantageous when they could be constructed in areas where the local streams moved slowly or where the flow of the streams was irregular and unpredictable. Tides are highly predictable and unlike streams are not dependent on rainfall. But the tides powered the mills only intermittently and at times that were not always convenient for the mills' operators. Because the length of the tidal cycle does not evenly divide the length of the day, the operating hours of the mill were constantly shifting, sometimes occurring during the day and sometimes late at night. On balance

Part of the Rance tidal power plant in Brittany, France. Completed in 1967, it remains by far the largest facility of its kind. *(La Rance Tidal Power Plant)*

the mills' operators must have felt that the benefits outweighed the costs, because tidal mills were operated throughout Roman times and some were built and operated throughout the European Middle Ages. The *Domesday Book,* for example, the 11th-century inventory of William the Conqueror's English holdings, lists at least one tidal mill.

Scaling up the simple-sounding tidal mill to create a commercially viable electricity generating station is no small challenge. Numerous proposals for harnessing tidal power were proffered throughout the first half of the 20th century. But constructing this kind of generating station is difficult because it involves undertaking a massive civil-engineering project in the middle of a large tidal basin. The problems associated with this type of project were first solved by French engineers with the successful construction of the La Rance Tidal Barrage, which is located in northern France. Construction began in 1961. It has operated successfully since 1967, and it remains the world's only commercial-scale tidal barrage.

At La Rance the difference in height between high and low tides averages about 28 feet (8.5 m). The tidal variation—the difference between high and low tide—changes in a predictable way from neap tide to spring tide, but the difference is always sufficient to generate power. A large tidal variation is, however, by itself insufficient to warrant the construction of a tidal power facility. To create a commercial-scale facility, it is also necessary to impound enormous amounts of water behind the barrage. In the case of La Rance, the basin behind the barrage has a capacity of about 6.4 billion cubic feet (180 million m^3) of water in an area of 8.5 square miles (22 km^2). The basin fills up and empties out twice each day. The La Rance Tidal Barrage works as follows:

A 2,460-foot (750 m) long barrage was built across the mouth of the basin effectively separating the basin from the ocean. (The section of the barrage containing the turbines is 1,000 feet [330 m] long.) As the tide rises, seawater flows from the ocean through gates in the barrage and into the basin behind it. At high tide the gates are closed, impounding the water behind the barrage. The operators wait for the tide to ebb in order to create a hydraulic head sufficient to drive the turbines. When there is enough difference between the water levels in front of and behind the barrage, the water is allowed to flow through the facility's 24 turbines, each with a rated capacity of 10 MW for a total capacity of 240 MW. As the water behind the barrage flows back to the sea, it turns the turbines, which drive the generators, which produce the electricity. Electricity production continues until the water level behind the barrage is reduced

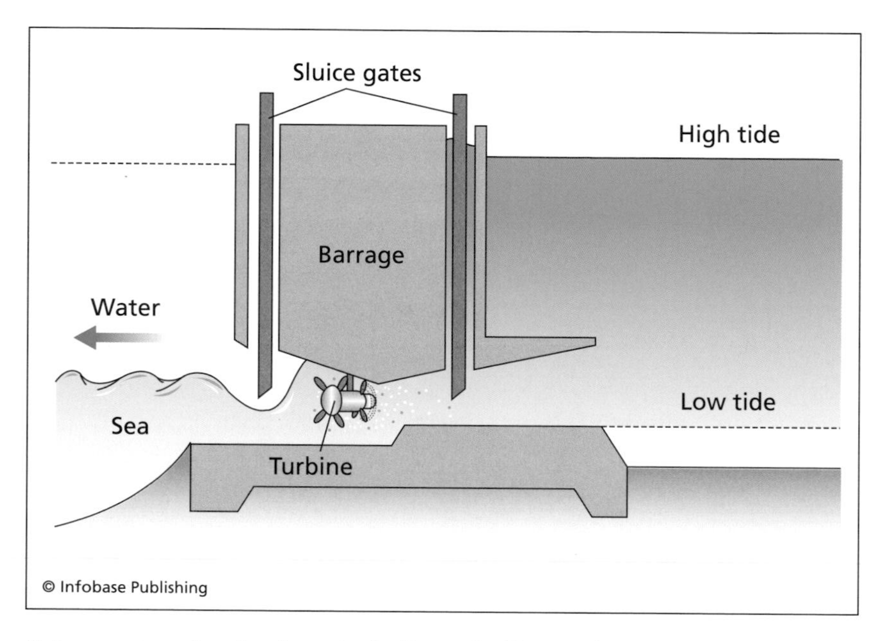

Cutaway view showing how the La Rance facility produces power.

to a level where power production is no longer practicable, at which point the sequence repeats.

This sequence of events can be modified in two different ways. First, La Rance can be operated so as to produce electricity as water flows into the basin as well as out. The incoming water is directed through the turbines, which are allowed to generate electricity on the incoming tide. Second, the La Rance facility can be operated in a way somewhat analogous to a pump storage facility. The station can pump water behind the dam in order to increase the height of the hydrodynamic head. As with all pump storage facilities this requires more electricity than is generated by the turbines as the impounded water returns to the sea, but as with all pump storage facilities, if the electricity required to pump the water is obtained during a period of low demand, and electricity production can be deferred to a period of high demand, the procedure can be justified on environmental and economic grounds. (See the sidebar "Pumped Storage" on page 30 for a description of this type of technology.)

The site that is always mentioned as a possible location for a La Rance–type facility is the Bay of Fundy in eastern Canada, the site of the world's highest tides. The maximal tidal variation in the Bay of Fundy is approximately 70 feet (21 m); the bay covers 3,600 square miles (9,300 km^2), and twice each day 70 billion cubic feet (2 billion m^3) of water flows into and out of the bay. The tidal variation varies from point to point within the bay, but the flow is high essentially everywhere. Engineers and planners have discussed building a tidal energy facility here since the 1920s. In 1984, Canada opened a small tidal barrage facility in the Bay of Fundy. It operates along the same principles as the La Rance facility but is in every way much more modest in scope. The Bay of Fundy facility has a capacity of 18 MW. An even more modest facility was constructed in Russia on the Kola Peninsula on the Barents Sea in 1968. It has a capacity of 0.4 MW.

As is apparent, tidal barrages share many of the characteristics of low-head hydropower facilities. In particular, a large tidal barrage facility has the potential to produce large amounts of electricity in a predictable way with zero emissions. Even better: Building a tidal barrage, though very expensive, carries far fewer social costs than building many large hydropower projects because the construction of the barrage cannot result in the displacement of a large number of people. A barrage must be built between the ocean and a large piece of land that is inundated by the ocean twice each day. No one lives in tidal basins, and so no one will be displaced by the barrage.

There are also characteristics to tidal barrages that are disadvantageous. Although a large tidal barrage can produce a significant amount of power each day, its output is determined more by the flow of the tides than by consumer demand. The La Rance facility, for example, produces enough power for 200,000 homes, but for part of each day it produces no power whatsoever. The development of this type of technology is further restricted by the scarcity of suitable sites, which require a large tidal variation and a large basin in which to impound the water for later release. Approximately 20 sites have been identified throughout the world for possible development, but some are in remote locations devoid of infrastructure or consumers. Tidal barrages can make a difference in the right area, but even if all identified sites were developed their total contribution to the world's electric power production would still be modest.

TURBINES WITHOUT DAMS

There is a second technology available for converting tidal energy into electricity. This technology may be more widely applicable than tidal barrage technology because it does not rely on barrages at all. Instead, it extracts energy directly from currents generated by the tides. Technically, this approach is new. Engineers have only recently begun to work on the problems associated with converting the energy of motion of ocean currents into electricity, but conceptually the idea is similar to the much older idea of the wind turbine. By placing a turbine in an ocean current, the linear motion of the water can be converted into rotary motion by a device similar to a wind turbine. Some of these undersea turbines even look like wind turbines, and as with wind turbines, the rotary motion of the undersea turbine can be harnessed to produce electricity. The ocean currents play the same role as air currents. Before engineers decided to begin searching for technical solutions to accomplish this goal they had to decide whether it was worth the effort—that is, they had to evaluate the potential of ocean currents as an energy source. Deploying these turbines is difficult and expensive, so is it worth it? How much energy can be extracted from an ocean current?

The waters of the ocean are in continual motion. Some currents occur far from shore; some are within a few miles of shore; some run continuously and some run intermittently. And with today's technology some can be exploited to produce power and some can-

SeaGen marine current turbine awaiting installation in Strangford Lough, Belfast

not. The costs associated with building and maintaining a deep-sea turbine farm, for example, are prohibitive. Consequently, most oceanic currents cannot be exploited for electrical power production at present. But currents that run near the shore are a different matter because they are close enough to land to offer at least the possibility of economical exploitation. These currents are heavily influenced by the topography of the coast and the topography of the ocean bottom, and many of these currents are driven, at least in part, by the tides. (The machines designed to convert the energy of tidal flows into electricity are called tidal mills.) The drawback of attempting to harness tidal currents is that they stop and reverse direction four times every 25 hours. Of course, when the currents stop, they fail to drive the tidal mills and electricity output drops to zero, but tidal currents are, at least, predictable, and some are reasonably swift. For example, in Uldolmok Strait in South Korea, a proposed location for a set of tidal mills, the tidal current has a top speed of about 20 feet per second (6 m/s). Tidal currents are not usually this fast, but in areas where there is a constriction in the channel along which these currents flow, the velocity increases as the cross-sectional area of the channel decreases. (The equation that relates the speed of the current to the cross-sectional area of the channel through which it passes is equation [4.1].) The best place to locate a mill is, therefore, in a narrow channel.

The maximum power that can be converted into electricity from an ocean current is best expressed in a simple algebraic equation. That equation is as follows:

$$P_{max} = \frac{CdAv^3}{2} \qquad (5.1)$$

where P_{max} represents the maximum power that can be converted from the current; d represents the density of the water, and v represents the speed of the flow before it encounters the tidal mill. (The mill slows the current as it converts some of the current's energy of motion into electricity.) The letter A represents the cross-sectional area swept out by the mill's rotors, and the letter C represents the maximum efficiency attainable by the mill—or to put it another way: C represents the maximum *percentage* of the current's power that can be converted into electrical power. (The quantity $dAv^3/2$ is the power of the current flowing through the circular area swept out by the blades of the tidal mill.)

Equation (5.1) explains why designers emphasize the importance of locations with faster currents. The speed in this equation is cubed. This means that power depends very strongly on speed—so, for example, if the speed of the current is doubled from v to $2v$, the power of the current jumps by a factor of 8 ($2v \times 2v \times 2v = 8v^3$). In particular, a five-mile per hour (8 kph) current possesses double the power of a four-mile per hour (6 kph) current. Therefore, even small increases in velocity can lead to substantial changes in the amount of available power. By contrast, A, the cross-sectional area of the flow passing through the mill's blades, has less of an effect on the mill's output in the sense that doubling the cross-sectional area only doubles the mill's output. The density of the water, *d,* is, of course, completely outside the control of the engineer, so with respect to producing the maximum possible power, it is important to build a tidal mill with the largest possible rotor and position it in the area with the fastest possible current.

During the 1920s at Göttingen University, the German scientist and engineer Albert Betz developed a theoretical description of how windmills worked and how much energy they could produce. Betz's reasoning, however, also applies to tidal mills. Equation (5.1) was an important part of his model. Betz's mathematical model predicted that *C* is no larger than 0.59—that is, no more than 59 percent of the power of a moving current—no matter whether it is an air current or a water current—can be converted into electricity. Because water is very dense, which makes *d* in equation (5.1) very large, the amount of power that can be harnessed from tidal mills is, in principle, also very large, even in a slowly moving current. (More recent research indicates that *C* is less 0.5, although its exact value remains an open question.)

Tidal mills come in a variety of shapes. The easiest to recognize are those that look like the more familiar wind turbines. Marine Current Turbines, a company based in the United Kingdom, began with a single 300 kW prototype with a rotor 36 feet (11 m) across leading to a cross-sectional area—represented by the letter *A* in equation (5.1)—of about 1,000 square feet (95 m^2). Because tidal currents drove this prototype, it was not capable of producing a steady current of 300 kW. As the tidal currents change direction they first slow to a stop. This happens four times every 25 hours. Electrical output drops accordingly, but the output does not remain at zero for long. As the current reverses direction, the turbine begins to produce electricity again—that is, power is produced as the tide rises and as it falls. Today (2009) Marine Current Turbines operates one larger dual turbine unit that produces 1 MW of power. As of

2009, contracts have been signed to deploy units off the coast of western Canada and Northern Ireland. Other designs very different from the Marine Current Turbine design have also been developed. One idea of particular interest incorporates a helical turbine, which in 2001 was patented by the Russian engineer Alexander M. Gorlov. (A helical turbine is mounted on a vertical axis and resembles an eggbeater.) Gorlov's turbine is said to be capable of converting 35 percent of the power of the current, which is quite good compared to other, more standard designs.

The different designs currently in place or under development have a number of commonalities so it is important not to overemphasize differences. All are located (or will be located) below the surface of the ocean in locations where the current runs relatively quickly. All designs convert a fraction of the energy of motion of the tidal currents into electrical energy and send it back to shore through a submarine cable. They produce no emissions. They do not require a barrage to function, and their environmental impact is less than that of a tidal barrage.

But it cannot be asserted that these devices have no environmental impact. It is not clear, for example, what effect these huge spinning blades will have on marine life, especially whales and other large marine organisms. Furthermore, a line of large tidal mills across the mouth of a bay would certainly change the flow rate of the tides. The environmental effects of diminishing the tidal currents in a particular location depend on considerations unique to each location. These facts simply serve to emphasize the fact that large-scale power production cannot be done without affecting the environment.

A final commonality worth discussing here is that all of these schemes are substantially more expensive than wind, which, as previously mentioned, remains more expensive than, for example, coal. Tidal mills will, presumably, eventually be subject to the same laws of the marketplace as other types of power generation technologies.

Spokespersons for the companies that manufacture these devices always stress that it is early yet, and that as production volumes and manufacturing experience increase, prices will come down. This is certainly true, but whether they come down enough to be competitive with other technologies without generous government subsidies is not clear at the present time.

Heat Engines

This chapter describes how heat from the ocean has been harnessed to do work. But before one can appreciate the advantages and disadvantages of this technology, it helps to know a little about heat engines in general.

THE THEORY OF HEAT ENGINES

Heat engines are machines that convert heat into work. The first heat engines, which were built more than 300 years ago, were steam engines used to drive water pumps. Powered by coal, they were used to pump water out of coal mines. In the 19th century, heat engines—also powered by coal—provided the motive force for trains and ships. Although the transportation sector still depends almost exclusively on heat engines—jet engines and internal combustion engines are both examples of this type of energy conversion technology—petroleum is now the fuel of choice. But in many countries,

Nicolas-Léonard-Sadi Carnot. He discovered the principles governing the operations of heat engines. *(AIP Emilio Segrè Visual Archives)*

coal-fired steam engines remain the preferred method for generating electricity. The United States, China, Germany, and India, for example, are all heavily reliant on coal technology to power their national grids. Some countries also use natural gas–fired heat engines to produce electricity, and some nations still find it necessary to burn oil to meet some of their electricity demands.

Fossil fuels are not, however, the only primary energy source used to supply heat engines with the thermal energy they need to operate. Nuclear power plants, which harness the energy within the nuclei of atoms, are also heat engines. In the United States nuclear power is responsible for about 20 percent of all electricity production. In France the figure is closer to 75 percent. Worldwide, most electricity is generated by heat engines of one type or another.

One characteristic shared by all of the different types of heat engines mentioned in the preceding paragraphs is that they have fairly high operating temperatures. The exact operating temperature for a particular engine depends upon the type of engine under consideration, the type of fuel it uses, and the way that the engine is operated. Nevertheless, all of these engines operate at temperatures that are high enough to be hazardous to humans.

But a high operating temperature is not a requirement for a heat engine because, strictly speaking, heat engines only require a temperature *difference* to operate. Just as waterwheels operate between two different water levels, heat engines operate between two regions of different temperatures. With respect to waterwheels, as water flows from the higher level to the lower level, the waterwheel converts some of the water's energy of motion into work. With respect to heat engines, heat will flow from a region of higher temperature to a region of a lower temperature, and as it flows, the engine converts some of that thermal energy into work. This analogy between waterwheels and heat engines was proposed by one of the 19th-century's most insightful thinkers, the French scientist, Nicolas-Léonard-Sadi Carnot (1796–1832) as he attempted to explain how heat engines generate power.

Carnot recognized that the efficiency of an engine is determined solely by the percentage of "flowing" thermal energy that is converted into work. An engine that converted all of the thermal energy that it produced into work, for example, would be 100 percent efficient. If it converted half of its thermal energy into work, it would be only 50 percent efficient. Carnot discovered that each engine has a maximal efficiency that it can attain, and that maximum value is always less than 100 percent. This upper limit on its efficiency does not depend on the details of the engine's construction or the fuel it uses. Nor does it depend on how the engine is operated. Instead, the maximal efficiency is determined solely by the temperatures of the two regions between which the engine operates. The lower temperature region is usually, but not always, taken to be the temperature of the outside environment; the higher temperature is the operating temperature of the engine. The bigger the temperature difference between these two regions, the more efficiently the engine can, in principle, operate. Again: Efficiency is the percentage of thermal energy moving from the higher temperature region to the lower temperature region that is converted into work.

This upper limit to an engine's efficiency is easily summarized by an algebraic equation:

$$E = \frac{T_h - T_l}{T_h} \qquad \textbf{(6.1)}$$

where T_h represents the (higher) operating temperature of the engine, and T_l represents the (lower) temperature of the environment, and both temperatures are measured in degrees kelvin. (A temperature difference of one degree kelvin equals a temperature difference of one degree Celsius, but 0°C equals 273.15°K.) The letter E represents the maximal efficiency of the engine. For any given upper temperature, T_h, and lower temperature, T_l, it is easy to make an engine that is less than maximally efficient, but it is impossible to make an engine operating between these two temperatures whose efficiency exceeds this upper limit.

To see how this works in practice, consider an engine whose operating temperature, measured in degrees kelvin, is 1,200°K (1,700°F, or about 927°C), and suppose that the temperature of the environment is 300°K (80°F, or about 27°C). This engine cannot convert more than 75 percent of the thermal energy produced within the engine into work (0.75 = (1200 - 300)/1200). By contrast, if the engine's operating temperature is 315°K (108°F or 42°C), and the temperature of the environment is 300°K (80°F or 27°C), the engine can, at best, operate at about 4.8 percent efficiency, which is another way of saying that it can convert no more than 4.8 percent of the available thermal energy into work; the rest, which in this case is 95.2 percent of the total, is "waste" in the sense that it cannot be converted into work.

Equation (6.1) indicates why engineers design heat engines that operate at temperatures that are as high as possible. An engine with a higher operating temperature can, in principle, convert more of its thermal energy into work. Therefore higher temperature engines will, in theory, waste less because they will require less thermal energy to produce the same amount of work. These ideas are extremely important when evaluating proposals to tap the enormous supplies of thermal energy contained in Earth's oceans.

PRACTICAL APPLICATIONS

A small fraction of the thermal energy of the world's oceans can, in principle, be converted into electricity by heat engines that operate between the warmer surface waters and the cooler depths. The machines designed for this purpose are collectively called ocean thermal energy converters (OTECs). The first proposal to build an OTEC dates back to 1881 when the French scientist Jacques-Arsène d'Arsonval (1851–1940) suggested harnessing the temperature difference between the warmer upper layers of the ocean and the cooler lower regions to do work. D'Arsonval was a creative scientist who was interested in both physics and biology and the interplay between the two fields. He was also an inspirational teacher. His student Georges Claude (1870–1960) was the first to attempt to deploy an OTEC device. He tried twice—the first time at Matanzas Bay, Cuba, and the second time in a cargo vessel moored near the Brazilian coast. These machines were purely experimental in character—they required more power to operate than they produced. Experience has shown that building a productive and economical OTEC plant is a difficult technical challenge even today.

At its most basic level, OTEC plants run on solar energy. The world's oceans act as enormous, if not especially efficient, solar collectors. As the Sun shines on the oceans, some of the light from the Sun is reflected at the surface, but enough of the Sun's energy is absorbed by the water to raise the temperature of the water in a predictable and potentially useful way. The effect, not surprisingly, is especially strong in tropical regions. Absorbed light raises the temperature of the water that absorbed it, and this increase in temperature causes a small increase in volume. As a consequence, water

These are parts of a machine for generating electricity based on the differences in temperature between the sea surface and great depth. Devised by Georges Claude in 1926, this device was a forerunner of modern ocean thermal energy conversion (OTEC) projects. *(Oceanographic Museum of Monaco and National Oceanic and Atmospheric Administration/ Department of Commerce)*

near the surface of the sea is slightly less dense than water deeper in the ocean; the warmer, less dense water floats on top of the colder, denser water, and there is very little mixing between the lower and upper regions. Because sunlight does not penetrate very deep into the ocean, the cooler water is "only" about 3,000 feet (1,000 m) from the warmer surface. In the tropics the temperature of the water near the surface of the ocean is about 77°F (25°C or 298°K). The lower layer has a temperature of about 40°F (5°C or 278°K).

As indicated in equation (6.1), heat engines operating between these two temperatures cannot hope to be more than about 7 percent efficient, and practical engines would operate at still lower

(continued on page 96)

Ocean Thermal Energy Conversion Tests

OTEC technology is slowly evolving, but even after decades of work only a few nations have deployed demonstration plants. Japan and the United States were the first to establish significant OTEC research programs. For about one year, beginning in 1982, a group of Japanese companies operated a demonstration plant on the island nation of Nauru. The plant produced about 100 kW of electricity, about two-thirds of which was used to run the plant's pumps and other equipment. The United States performed its OTEC experiments in Hawaii, the best known of which was a demonstration plant that operated from 1993 until 1998. It produced up to 225 kW, about 60 percent of which was used to run the plant.

These plants tested OTEC designs, and the results were not entirely satisfactory and did not lead to commercial-scale plants. The United States Department of Energy (DOE) even ceased funding research into OTEC.

During the early years of the 21st century, India built a one-MW OTEC demonstration plant. But that project suffered from cost overruns, and the plant, which was designed to operate on the open ocean, suffered damage in rough seas. Funding was canceled. Yet interest in developing OTEC technology also continues in India.

The attraction of OTEC rests on two observations. The first is easy to appreciate: The fuel, which consists of warm water, is free. The second reason OTEC continues to attract attention depends on the observation that there is an enormous amount of energy available in the temperature difference between the upper and lower regions of tropical oceans—enough energy, in theory, to power a nation. And yet not one of the three countries that invested in demonstration plants has attempted to build a commercial-scale OTEC power plant. By contrast, for many years, all three nations have successfully operated commercial nuclear power plants, which to many observers may appear to be a technologically more difficult undertaking. Why not at least one OTEC power plant?

OTEC power plants are, for several reasons, much harder to build than, for example, a nuclear power plant. First, because an OTEC plant operates between thermal reservoirs that are at almost the same temperature, they are, under the best of circumstances, able to convert into work no more than 6 or 7 percent of the thermal energy that they produce. (Recall that the larger the temperature difference, the larger the percentage of thermal energy that can, in theory, be converted into work.) As a consequence, OTEC plants must operate very near their optimum level of efficiency in order to accomplish anything at all. By contrast, heat engines that operate between larger temperature drops have a larger percentage of energy that they can convert into work. For high temperature engines, inefficiencies are less of a problem. (Early steam engines, for example, wasted almost all of the thermal energy they produced yet usually managed to perform the functions for which they were designed.) Second, as an OTEC plant grows in size, its power output increases faster than its own internal power requirements. A small OTEC plant will consume more power than it produces, while a large plant will consume considerably less. Consequently, the only profitable OTEC plant would be a very large OTEC plant, and the larger the better. This increases the risk to investors, who must decide between building a large and expensive plant or none at all. Investment risk would be especially high during the first few generations of OTEC plant design, when engineers and operators would still be learning how to make such a plant profitable. Finally, every OTEC plant operates with some—or in the case of the Indian plant, all—of its equipment in or on the sea, an environment that is corrosive and sometimes violent. It is a difficult place in which to locate a large power plant of any sort.

With respect to OTEC technology there is a huge difference between theory and practice. It is not known where or when the first commercial-scale OTEC facility will be built.

(continued from page 93)
efficiencies. On the one hand, one could say that such low efficiencies are terrible, and an engine that is able to utilize no more than 7 percent of the heat flowing through it is hardly worth developing. For many applications, this is certainly true, but the oceans are so huge and hold so much thermal energy that even if one were only able to economically harness a tiny fraction of their thermal reserves the implications would be enormous. And, of course, the fuel, solar energy, is free and abundant.

Several different designs for OTECs have been proposed, and various small-scale prototypes have already been built but with limited success. One representative design will be considered here. It works as follows: A very long, large pipe extends deep into the ocean in order to draw cool water from the depths into the power plant. Another pipe is used to draw warm surface water into the plant. Inside the plant, situated between the supply of cool water and the supply of warm water, is a third material called the working fluid. The working fluid is chosen so that it will boil when warmed by the warm water and condense when cooled by the cool water. The working fluid does not come into physical contact with either water source. Instead, only heat is exchanged between water and working fluid via radiator-like devices called heat exchangers.

First the working fluid acquires heat from the warmer water via a heat exchanger. As its temperature increases, the working fluid turns to vapor. The expanding vapor is directed against the blades of a specially designed turbine, causing it to spin. The turbine is connected to a generator, which produces electricity. Once past the turbine, the working fluid exchanges heat with the cooler water—again through a heat exchanger—this time giving up some of the heat it had earlier acquired from the warmer water. As the working fluid cools, it condenses. The working fluid is then pumped back to the first heat exchanger in order to acquire heat from fresh warm water. The cycle repeats.

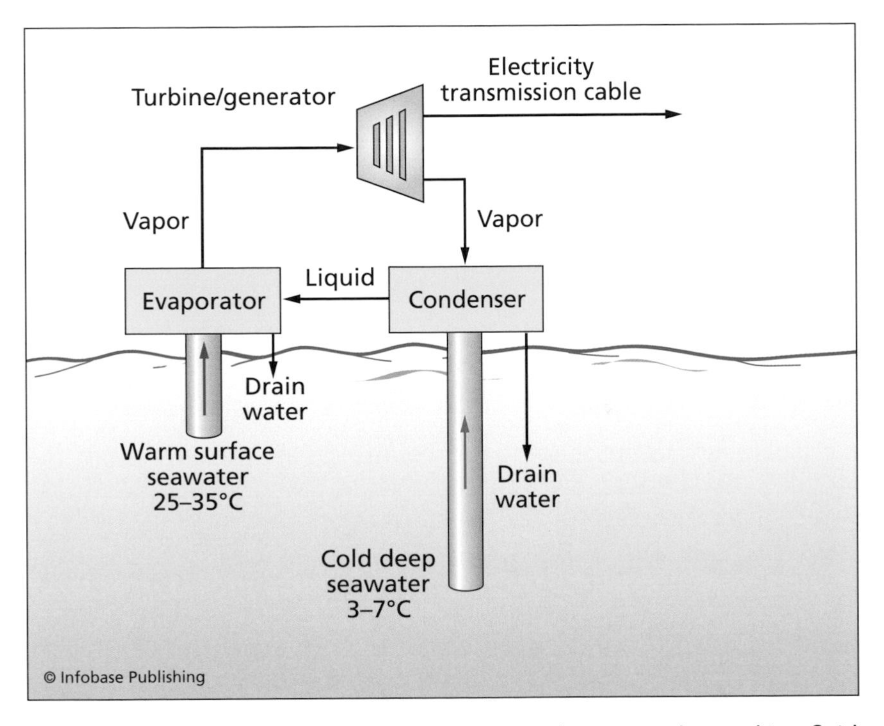

OTEC process diagram. The heat exchanger used to warm the working fluid is called the evaporator. The heat exchanger used to cool the working fluid is called the condenser.

There are a number of technical problems associated with building a large-scale OTEC generating station, but the fundamental problem is the low efficiency of the unit. Because a practical OTEC plant will only be able to convert into work a very small percentage of the heat carried by the warm water, it requires enormous volumes of warm water to produce useful amounts of electricity. And the same statements hold true of the cool water: Very large volumes of cool water must be brought up from the deep to cool the working fluid. Large-scale OTEC plants will require huge pipes, huge pumps, and huge heat exchangers to facilitate the rapid transfer of significant amounts of heat. Everything will have to operate very efficiently in order to produce commercially meaningful amounts of electricity, and this raises costs and additional technical issues. Furthermore, the energy needed to operate on this scale must be drawn from the power output of the plant. Only what is left after the plant's energy needs have been met is available for sale.

The energy crises that occurred in the United States and other Western nations during the 1970s as the result of rapid oil price increases caused researchers in the United States and elsewhere to focus on the possibility of developing OTEC plants. (At the time, it was common practice to burn oil to produce electricity.) Research into OTEC continued throughout the late 1970s and early 1980s. OTEC technologies were attractive to the researchers of the time because the "fuel"—solar energy—was perceived as free and a viable alternative to high-priced oil. OTEC technology offered the promise of enormous quantities of electricity for as long as the Sun shone. But the technical problems encountered by engineers attempting to produce a working, economically competitive OTEC plant were insurmountable using 1970s technology. As the price of petroleum stabilized and began to diminish, government support for OTEC research evaporated. In 1984 the U.S. Department of Energy spent just $8.2 million on OTEC research, less than one-tenth of what it had spent on OTEC in 1978.

Today, the prices of natural gas and oil have once again become volatile. Concerns about the effects of greenhouse gas emissions have further focused the attention of government research facilities on finding less expensive and less polluting technologies with more predictable costs. Some engineers are renewing efforts to develop OTEC technologies. The initial market would be tropical islands, locations where fossil fuels must be imported at a premium and where the temperature difference between the ocean's upper and lower layers is stable and comparatively large. The long-term goal, however, is the harnessing of the oceans' temperature difference to produce electricity on a scale sufficient to supply larger markets. The day that OTEC becomes commercially viable—if that day ever arrives—is still many years in the future.

7 The Role of Government in Promoting New Technologies

Every power generation technology described in this book is (or has been) heavily dependent on government subsidies. Large-scale conventional hydropower has often involved the damming of rivers, the flooding of valleys, and sometimes the forced displacement of many people, all activities that are difficult to accomplish without enthusiastic government support. Once a project is completed, however, conventional hydropower is dependable and often competitive with all other technologies. (Although many hydropower facilities could be operated profitably, sometimes, as discussed in chapter 3, governments decide to continue to subsidize the electricity produced at these facilities in order to spur economic development.) There are other technologies—wind is the most prominent example—that require subsidies if the facilities are to be built, and they continue to require subsidies even after a project is completed because the price of electricity produced by these power

U.S. Department of Energy headquarters *(DOE)*

sources is not yet competitive with that of other more conventional power sources.

To be fair, there are many other countries where conventional power sources—a term that is shorthand for coal, natural gas, conventional hydropower, and nuclear, the technologies that produce the vast bulk of the world's electricity—also receive government subsidies of various sorts. In the United States, for example, where roughly half of all the nation's electricity is produced by burning coal, many experts argue that the full costs of burning coal are not reflected in the price U.S. consumers are charged for the electricity these plants produce. These costs include but are not limited to the environmental and social costs of extracting coal as well as the significant environmental costs associated with burning it. The failure to create a system that takes into account the environmental and social costs associated with coal use, they argue, is the most significant governmental *subsidy*. They argue that the government should create a price structure for electricity produced by coal-burning plants that would better reflect the actual costs of that technology. At present, the owners of coal plants retain the profits for themselves, but they pass along most of the environmental costs for all to share. That, at least, is the perception of many. Nations have found it difficult to act on this perception, however, because there is no generally agreed upon method of assessing environmental costs, and because the inexpensive and reliable electricity coal-burning plants now produce is so beneficial to so many.

And what is true of coal is, in fact, true of every energy production technology. How should the value of inexpensive, reliable electricity production be assessed relative to the values of reduced greenhouse gas emissions, energy security, and so on? What is the value of any power-production technology relative to any other power-production technology? And how can energy markets be designed that incorporate these values? There are no generally agreed upon answers to any of these questions.

Despite a lack of consensus on the details, the case for power-production technologies that enhance energy security and reduce greenhouse gas emissions is strong enough that many governments have already undertaken programs to encourage the immediate deployment of alternative technologies. But in the current energy markets, investors have usually shown little interest in implementing these alternatives without substantial government subsidies. Some governments have, therefore, created policies to stimulate interest by essentially guaranteeing the profits of those who invest in government-sanctioned methods of electricity production. The results of these governmental programs have varied widely. It is, therefore, worthwhile to see what types of programs have worked and what types have not. But rather than try to account for all of the programs for each of the technologies described in this volume, the emphasis in this chapter is on the strategies used by nations to encourage the deployment of wind power. Because with the exception of conventional hydropower, wind is the most widely deployed of the technologies described in this volume, and it has long benefited from the largesse of several countries. More recently, some governments have begun to apply wind power–like subsidies to encourage the development of other power production technologies, including those technologies described in the preceding chapters. The policies of the United States, Denmark, and Germany, three of the world's biggest wind-power producers, are of most interest. (The history, technology, and environmental implications of wind are discussed in chapters 8, 9, and 10, respectively.)

THE UNITED STATES: CREATING SUPPLY AND DEMAND

In the United States, the federal government supports wind energy in three ways: direct subsidies, regulations to encourage wind energy, and taxpayer-subsidized research and development into wind-energy technologies. U.S. attempts to develop and deploy wind-energy technology began in earnest with the energy crises of the 1970s, which were marked by rapid increases in the worldwide cost of oil and, in 1973, a brief embargo on oil shipments to the United States by some oil exporters as retaliation against U.S. support for Israel during the 1973 war between Israel and several of its neighbors. (Within the United States, some states also developed their own programs to encourage the development of wind energy. Most notable among these are the efforts of the state of California. In fact, so successful was California in developing wind capacity that during the 1980s most of the electricity produced via wind energy throughout the world was produced within California, despite the fact that most of the best U.S. sites for wind farms are located in states other than California.)

Important early federal support for wind energy was contained in the Public Utility Regulatory Policies Act (PURPA) of 1978. PURPA, which was part of the National Energy Act of 1978, required utilities to buy electricity from nonutility power producers that produced electricity using renewable energy technologies, provided that the capacity of these facilities did not exceed 80 MW. In a general way the law even provided guidelines as to the price that the utilities had to pay: Utilities were required to buy the power at a rate equal to the amount of money that they would have spent producing the power themselves. This was called the "avoided cost" of the power. Avoided cost was, however, a flexible standard and different states interpreted the standard in different ways. California and New York interpreted the standard in ways that favored the independent power producers, while other states used interpretations that favored the utilities. (The differences among states in the definition of avoided costs were large, and in 1995 the Federal Energy Regulatory Commission established national standards. These standards required that some states lower the rates at which utilities were required to purchase renewable energy, and some states were required to raise the rates that their utilities paid.) But when the United States passed PURPA, wind energy technology was still too primitive—in the sense that electricity from the wind was too expensive to produce—to flourish with this level of support. Wind technology of the 1980s was too inefficient to attract investors even with the federal subsidy. Large-scale wind farms were only found in California, which adopted additional subsidies generous enough to attract investors for wind farms that produced only modest amounts of very expensive electricity.

In 1992, Congress passed the Energy Policy Act (EPAct), which provided a subsidy in the form of a *production credit* of 1.5 cents per kilowatt-hour of power produced during the first 10 years of operation of each wind turbine. Private companies received a production tax credit—that is, money off their tax bills—and public power producers received a similar sort of subsidy called a renewable energy production incentive. (The credit has been revised upward and in 2009 is worth 2.1 cents per kilowatt-hour.) After a slow start, the credit has had an important effect on the installation of wind power capacity. The effect is most fully illustrated by what has happened each time this subsidy has been allowed to expire. This has occurred in the years 1999, 2001, and 2003. In each of the years following the expiration of the credit—that is, in 2000, 2002, and 2004—installation of new wind facilities plummeted. When the subsidy was reestablished, the installation of new facilities surged. To get an idea of the scale of this one program, consider that in 2005 the production credit for wind-power producers amounted to $330 million, and the Congressional Joint Committee on Taxation estimates that maintaining the credit will cost $2.8 billion from 2005 until 2015. EPAct has had the most effect of all direct federal subsidies aimed at U.S. wind power producers. Key provisions of EPAct, which expired at the end of 2008, were retained in the American Recovery and Reinvestment Act of 2009 including heavy subsidies for wind power.

The final type of subsidy considered here is government-sponsored research. The federal government has conducted research into wind-power technology for several decades. Much of this research has been conducted within the Department of Energy (DOE). Research into wind energy began in the 1970s, a time when the U.S. intensively studied many so-called nonconventional energy sources. In 1979, the DOE spent almost $60 million on wind-power research. But a few years later the price of petroleum declined and so did federal support for wind power research. By 1982, only $16.6 million was allocated within the DOE for wind-power research. Despite the sporadic nature of the funding, engineers have been very successful in lowering the cost of converting wind energy to electrical energy—from roughly 50 cents per kilowatt-hour in the 1970s to about five cents per kilowatt-hour today, a tremendous accomplishment.

Wind-power research continues to attract the attention of the federal government. Research occurs in three ways. First, the DOE conducts its own program of research; second, the DOE sponsors university research; and third, DOE researchers form partnerships with industry and academia. More generally, the goal of DOE programs is to pursue research that would probably not occur—or at least it would not be conducted with the same intensity—if the research were conducted solely by private industry. It is worth noting that although the DOE negotiates royalty-sharing agreements with industry partners in the event that a particular research program leads to a commercial success, the DOE has not required renewable energy companies to pay any royalties when joint research has led to profitable inventions. DOE research subsidies have created a fast-growing industry but one that remains highly dependent on various forms of government support.

The principal barriers to the development of wind farms within the United States are not directly addressed by federal programs. The first barrier to the development of some sites is local opposition. The reasons expressed for opposition to the development of particular sites are varied, but many of the objections are surrogates for the perceived ugliness of wind turbines. (See chapter 10 for a description of the opposition to the development of the United States' first offshore wind-power site.) The federal government has no policy to deal with this type of opposition, but other nations have developed strategies to deal with local objections. See, for example, the Danish model described in the next section.

Another important difficulty in developing more remote wind-power sites is associated with the difficulty of constructing high-voltage transmission lines to connect producers with consumers. Wind farms require large tracts of unpopulated land on which to build the turbines. These sites must be free from trees and buildings and are often found far from the cities that are their main markets.

(continued on page 110)

An Interview with Dr. Stan Bull on Research at the NREL

Dr. Stanley R. (Stan) Bull is the Associate Director for Science and Technology for the National Renewable Energy Laboratory, and in this capacity he directs the research and development programs at the Laboratory. He is also Vice President of the Midwest Research Institute. An accomplished scientist—he has authored approximately 100 papers—of more than 35 years, he has research experience in energy and related applications including renewable energy, energy efficiency, bioenergy, medical systems, nondestructive testing, and transportation systems, all in addition to his work of planning and evaluating technical programs. Here, he shares his ideas about the processes by which research projects are chosen and evaluated, processes that are critical for scientific progress. (This interview took place July 9, 2007.)

Stan Bull *(Stan Bull)*

Q: The National Renewable Energy Laboratory [NREL] is involved in research into a wide variety of energy conversion technologies—

A: Correct.

Q: For example, wind, hydropower, geothermal, biomass, photovoltaic. But NREL doesn't concern itself with OTEC [ocean thermal energy conversion], and I don't think that NREL has ever investigated wave-energy conversion technologies.

A: Let me clarify that. First, we don't research hydroelectric. We have not done research

in that area. We are funded by the U.S. Department of Energy [DOE]. So what we do is dictated by the programs that are funded at the Department of Energy (DOE). In the '80s we were involved in ocean research—primarily ocean thermal [energy conversion]—although we were involved in wave energy and other forms of ocean energy. But in terms of what we did in the laboratory, it was primarily ocean thermal. That research ended because eventually DOE and Congress chose not to fund that program any longer as funding for renewable energy was reduced. For a good long while—approximately 15 years—we have not done research into ocean technologies. We are now starting an ocean program again. There is a modest amount of money for that this year, and that would include both wave and underwater-current research, but not ocean thermal.

Q: How are these technologies chosen? What is the mechanism?

A: This a hard question to answer.

We do what the Department of Energy funds us to do. The Department of Energy does the research that Congress appropriates funds for them to do. What does the federal process look like, and how do we play in that? We support the Department of Energy in developing multiyear program plans. There are a variety of plans. There are things called roadmaps, multiyear plans, and annual plans. These are reviewed by outside experts. Once a program gets going, there are generally incremental changes to it. Things may happen that cause the increments to be bigger or sometimes smaller.

There is the annual budget process that Congress goes through and that federal agencies and the administration go through with Congress. They submit a prepared budget and that gets submitted to Congress in the State of the Union every year—you know all this, I'm sure.

Q: Sure.

(continues)

(continued)

A: That's when the budget process is rolled out. Congress has hearings that involve the Department of Energy folks and lab people like us—some of our people and myself are invited to testify—and eventually Congress appropriates, and generally it's with a fair amount of definition about where the dollars are appropriated. Those dollars go to the Department of Energy, and the Department of Energy makes decisions on what fraction goes to our laboratory and what fraction goes out by way of solicitations, where they have private industries compete for awards or universities compete. One of our goals is to support the Department of Energy in an integrating role so that we can be an objective adviser to them and give them good, sound technical input relative to the direction that the program should go and what the research priorities should be. It's a very complex enterprise that is in play.

Q: The Department of Energy began research into wind power in the 1970s—

A: That's correct. A lot of research priorities are driven by world events—the world oil crisis in 1973 drove the U.S. government to seek alternatives. You've also seen that phenomenon in more recent times when the price of oil went up into the $70 per barrel range. One of the major points of emphasis of the Congress and the administration today is, "What are some ways to displace our need for foreign oil?" Or oil in general—but our imports are larger than our domestic production today. And so the major emphasis is to determine some of the alternatives, some of the major alternatives. How do we get there? How do we accelerate that? So things like the price of oil drives additional investment in renewable energy—in alternative energy.

A vitally important question is energy efficiency. How can you use our energy sources efficiently so that you don't need to have more and more of it?

Q: The researchers in the Department of Energy—to go back to wind power as an example—reduced the price of a kilowatt-hour of electricity

from the time that they started in the 1970s until today by about an order of magnitude—

A: That's a good summary. The price has been reduced by R & D, and with the help of industry development, by approximately a factor of 10.

Q: How difficult is it to maintain a research effort over such a long time frame?

A: It's a challenge. One of the most important things with a research effort is that it be steady, not cyclic. It takes a while to build a capability. The human resource, the people, are the most important part of that. It takes awhile to find the best people in this business, to attract them, and get them into a situation with equipment and facilities where they can be effective. So if the budget goes up and down, the best talent are the first to find alternatives and leave. So you lose the strength of some of your best capability over time. So the most important thing is a steady budget.

But as you start to have success, you begin to hear the mantra that "this is a mature technology, so we can reduce our R & D investment." If you think about it, the U.S. government still invests in coal technology, and how old is coal technology? It's very old. Nuclear energy is more than 50 years old, and we continue to invest in that. So these technologies at NREL, while they may be maturing, they still can mature substantially more—

Q: When NREL undertakes a research project with such a long time frame, how do you judge success in what is, from a practical point of view, almost open-ended research?

A: First of all, virtually all of our research is applied research. Actually, I would call it technology development. Sometimes we say research and development, sometimes we use the phrase research, development, and demonstration, and sometimes we say research, development, demonstration, and deployment, because we partner and work very closely with

(continues)

(continued)

industry. Everything we do is intended for industry. We don't do anything that we believe will have an outcome not intended for industry. So our ultimate measure of success is, "Has it been useful to industry?" And, of course, not everything we do ends up being adopted by industry, because you do go down some blind alleys. You do go places that don't work out, but you learn from those experiences. That puts you on a different path, and you go down that path and are successful. But we judge our success by, "Was it useful for industry?" and "Is it going to be useful for industry?" and "Did it eventually become so?"

Sometimes it takes longer than you think it should. We'll develop a technology, and it doesn't seem to go anywhere, but then 10 years later there is another improvement. Something else facilitates or enables that. Or the price of oil doubles. There are a lot of factors that can ultimately help make it successful.

Q: On balance, how satisfied are you with the mechanisms by which research projects are chosen and pursued at NREL?

A: That's a hard question for me to answer. I would say that we are not always entirely happy with all of the directions that we need to follow, but over time we are able to adjust our research portfolio such that we really do believe that frankly we are on a pretty good track most of the time.

Q: I very much appreciate your time and your expertise. Thank you.

(continued from page 105)

It is, therefore, necessary to build corridors for high-voltage transmission lines to connect power producers with the markets that they serve. These high-voltage lines are often controversial because their construction requires that many individuals surrender some of their property to make construction possible. If the high-voltage lines cannot be constructed there is little reason to develop more wind farms, as the producers will have no way to bring their product to market.

DENMARK: CREATING SUPPLY, DEMAND, AND GOODWILL

Denmark is a country of about 16,639 square miles (43,094 km^2) and 5.3 million people, and it has long been a pioneer in wind power. Some of the earliest research into wind turbines was conducted in Denmark by Poul la Cour (1846–1908), whose contributions are summarized in chapter 8. Today, Danish companies are some of the most successful in the wind-power market, and Danish-built turbines can be found in countries around the world. Popular support for renewable energy, especially wind, is very strong in Denmark. This enthusiasm is reflected in government policies, and it is even fair to say that some of the enthusiasm is a result of those policies.

Given the right weather, Denmark can generate as much as 25 percent of its electric power needs with nonconventional power sources, the highest proportion of any developed nation. Most of this power comes from the wind. It is worth noting, however, that fossil fuel plants continue to comprise about three-fourths of Denmark's generating capacity.

Denmark began passing legislation to encourage the development of renewable power during the 1970s in response to the oil crises of that time. In 1979, the government established a program to encourage investment into a variety of renewable technologies, including wind. The incentive was a 30 percent investment subsidy, a program designed to reduce the cost of installing wind turbines. The investment credit diminished steadily over the following 10 years, and in 1989, when it was at 10 percent, it was eliminated. In 1981, Denmark began offering a production subsidy to reward the production of electrical power rather than the construction of turbines. By 1992, the motivation for renewable energy subsidies had changed from energy security to the desire to lessen the environmental impact of electricity production. This change in motivation was reflected in two pieces of legislation that were passed in that year. First, Denmark instituted a carbon tax—that is, fossil fuel technologies, which produce greenhouse gases as a by-product, were taxed according to the amount of carbon they vented to the atmosphere—and so the Danish government made the cost of wind energy more competitive relative to fossil fuel technology by making fossil fuel technology more expensive relative to wind. Second, Denmark required utilities to purchase renewable energy at a premium price, a price that was higher than what they would have paid for fossil fuel–based electricity. Finally, Denmark established a production credit, a direct payment from government to producer of about 2.8 cents per kilowatt-hour. This payment was gradually reduced, but when these measures were first instituted, independent producers of wind energy benefited from a package of subsidies that totaled approximately 4.4 cents per kilowatt-hour. Compared to the United States, this was an extremely generous set of subsidies.

The Danish government also promoted the creation of a wind turbine manufacturing industry. In 1990, the government guaranteed long-term financing of large-scale wind projects provided that the projects used Danish manufactured turbines.

Unlike the United States, which has no real mechanism for enfranchising potential local opposition to the siting of wind farms, Denmark has worked hard to make wind power attractive to its citizenry. This was accomplished in two ways. First, beginning in 1994, all municipalities were required to create plans that would include possible sites for the installation of wind turbines. The law did not require each municipality to site a given number of turbines, but the legislation did help to distribute among the broader population the burden of accommodating wind turbines while simultaneously allowing local governments a voice in how wind energy would be implemented within their jurisdictions. In addition to this directive, Denmark also passed legislation that facilitated the formation of wind-turbine cooperatives. Early legislation encouraged many small investors to buy shares in local wind projects so that the benefits of wind-power production could be shared among those who also bore the costs.

In the intervening years, all of the legislation for subsidies, cooperatives, financing guarantees, and so forth has been continually reviewed and regularly revised in a way that is gradual and predictable. This makes for a stable business climate, and stability enables investors to successfully plan for the long term. The cumulative effect of this effort has been to foster investment in wind power. Wind power in Denmark has grown rapidly and steadily for the last two decades.

But success also involves overcoming technical challenges. Specifically, there are difficulties in integrating an intermittent power source such as wind at the scale that Denmark has accomplished. Because wind power is currently used to meet about 20 percent of Danish electrical power needs—and that percentage will continue to rise—the intermittent nature of wind can cause problems in terms of balancing supply and demand. When the wind blows over large areas of Denmark, a great deal of power is generated. But if the wind fails over a large area, the Danes must find a way to cover a very significant shortfall in national electricity production. This is the challenge.

Denmark has responded by improving grid connections with its neighbors, Sweden and Norway, neither of which is dependent on wind power. (Norway's electricity is produced primarily through hydroelectric facilities, and Sweden depends in roughly equal measure on nuclear power and hydroelectric facilities.) These grid connections enable Denmark to balance supply and demand. Because Denmark is small, has substantial wind resources, and is located close to energy-rich neighbors, Denmark has found an answer to its energy needs that works well *for Denmark*. The Danish model may not be a suitable approach for a larger or more geographically isolated economy.

GERMANY: RAPID GROWTH, AMBITIOUS GOALS

Germany has significant wind resources. Currently, it is heavily dependent on coal, a fuel that it has in abundance, and nuclear power. But the German government has pledged to gradually shut down all nuclear facilities and to build no more. It has also pledged to minimize coal consumption in order to reduce greenhouse gas emissions. Simultaneously accomplishing both goals would be an extraordinary technical accomplishment. Germany's challenges are further compounded by the fact that much of its natural gas supply originates in Russia, and there are concerns that Russia will use its supplies of natural gas to gain political advantage over adversaries, as some claim it has already done with its neighbors Ukraine and Belarus. Reluctant to use natural gas, determined to minimize the use of coal, and having decided to eliminate nuclear power, and with little in the way of additional undeveloped conventional hydropower resources, Germany turned to wind as its best bet for future energy production.

It is important to bear in mind that in comparison to Denmark, Germany's economy is enormous. Creating a wind-energy sector capable of displacing much of the commercial nuclear output of one of the world's largest economies is a daunting challenge. Germany already produces more of its energy from wind than any other nation. In 2008 there were approximately 20,000 operating turbines producing about seven percent of its energy needs. Planners hope wind's contribution to the nation's electrical energy output will be about 20 percent by 2020. Because Germany pursues a strategy of aggressive electrical energy conservation, the growth in energy demand is small—less than 1 percent per year. Renewables are expected to grow at a rate of about 2.6 percent per year, so wind is on track to gradually displace other more conventional sources of energy. But as with Denmark, the growth of the wind sector depends heavily on government involvement in the energy markets.

As with Denmark and the United States, German interest in renewable energy dates to the oil shocks of the 1970s. At first, there was increased investment in coal, nuclear, and the so-called renewable forms of energy. (Motivations have changed over the intervening decades, and now the German government subsidizes renewable energy principally for the perceived environmental advantages.) Beginning in 1975 and continuing into the 1980s, Germany initially concentrated on wind turbine research.

In 1986, the German federal government began subsidizing the construction and operation of wind turbines. As designs were evaluated and experience gained, subsidies began to increase. By 1989, participants accepted into a special program had the option of choosing between a production subsidy—the government paid 4.3 cents per kilowatt hour—or a 60 percent grant to pay for the construction of wind facilities. In 1991, the German government passed its so-called Feed-in Law, which required utilities to purchase renewable power at 90 percent of the retail cost of that power. This requirement is similar in concept to the United States' PURPA legislation described earlier. All of the German subsidies were, however, more generous than those provided by the federal government of the United States. And in 2000, Germany instituted the Renewable Energy Law, which stated that a new wind-turbine project would be subsidized at a rate of 11 cents per kilowatt-hour for the first five years of production, at which point the subsidy would begin to decrease. This was a tremendous boon to independent power producers, and construction surged ahead.

Germany maintains a very robust wind-energy sector, but as is the case in the United States, the German wind-turbine sector remains dependent upon the generosity of the government. This means that as the industry grows so will the subsidies it requires. It is even possible that the subsidies will grow faster than the capacity of the industry to generate power, because as less profitable sites are developed, larger subsidies may be necessary to make them attractive to investors. Technically, Germany can continue to expand its wind capacity, but whether its populace will be willing to pay the ever-increasing subsidies necessary to make this possible is not clear. There are already some indications that support for wind power has begun to weaken.

PART III

Wind Power

Wind Power: A Brief History

Successful attempts to harness the power of moving air began long ago. Wind power enabled early pharaohs to sail the Nile. Windmills have been used for well over 1,000 years to pump water and grind grain. For centuries European explorers depended upon the wind to traverse the world's oceans. During the latter years of the 19th century, however, interest in wind power waned. Wind technologies, some of which were already fairly advanced, were displaced by fossil fuel technologies. Fossil fuels were easy to obtain, relatively easy to transport, inexpensive, and most importantly, fossil fuel–powered engines did not depend on the vagaries of the weather.

Even today all of these statements are still true of coal, but the situation has changed for oil and natural gas. Oil, once widely used as fuel for power plants, is rarely used to generate electricity anymore; its price is too volatile. Coal-fired and natural gas–fired

At the USDA-ARS Conservation and Production Research Laboratory, in Bushland, Texas, wind turbines generate power for submersible electric water pumps that are far more efficient than traditional windmills. *(Scott Bauer, USDA)*

power plants are as important as ever—in the United States, they provide about 70 percent of the nation's electricity—but they are often used in different ways. Coal-fired plants are usually operated for prolonged periods at fairly constant outputs. They do not compete with wind because they do not serve the same function. Natural gas–fired plants, by contrast, are often used intermittently, being turned on and off as the demand for power rises and falls. The situation with respect to natural gas is especially relevant to this chapter, because both gas-fired power plants and wind turbines are used to produce peak power. Natural gas–fired plants, though functionally very reliable, are often used intermittently because their fuel has become too expensive to burn continuously. Wind turbines are used intermittently because the wind only blows some of the time. But if the two types of power generating technologies are used in roughly similar ways, their cost structures are very different. Because new deposits of natural gas tend to be expensive to produce, and because natural gas supplies are increasingly prone to disruption, natural gas prices have become volatile, making long-term planning difficult for residential and industrial customers alike. By contrast, wind energy has a well-understood cost structure and although intermittent in supply, its physical characteristics, especially its average rate of supply, are predictable. Wind, once forgotten, has again become one of the fastest growing sectors of the power-generation market.

In addition to changes in the economics of fossil fuels, the environmental costs associated with a heavy reliance on fossil fuels are becoming increasingly apparent. The consumption of fossil fuels results in the release of various pollutants. Some of the pollutants are local in nature and contribute to smog or acid rain, for example. Carbon dioxide emissions, however, are much longer lasting, and because they affect the global climate, burning fossil fuels could not have more widespread environmental implications. Wind turbines, as will be seen, also have environmental effects. These effects are not especially obvious only because wind farms generate relatively little power at present. If wind becomes a large-scale source of electricity, its effects will generate more controversy. Nevertheless, when compared to fossil fuel–fired power plants, a heavy reliance on wind has far fewer environmental implications.

Today, engineers and scientists are attempting to make the maximum possible use of this resource, which is inexhaustible and emissions-free. This chapter provides a brief history of the evolution of wind-power technology.

WINDMILLS

Not much is known about the very early history of windmills. They seem to have been invented in Persia, which was located in present-day Iran, or Mesopotamia, which was located in present-day Iraq. Some claim windmills first appeared in China. The earliest written references are, however, Persian, and date to the seventh century C.E. These early windmills were used to mill grain and possibly to pump water. But if there is uncertainty about their origin and initial application, there is, at least, general agreement on the basic elements of their design: The arms of ancient windmills were mounted on a vertical shaft so that they turned in a plane that was parallel to the ground, and they were somewhat similar in appearance to revolving doors. The *sails* of these windmills were often made of reeds or wood, and as with modern revolving doors, these early windmills were also partially enclosed by walls.

If building a windmill indoors seems counterintuitive, imagine what would happen if the windmill were not partially enclosed by

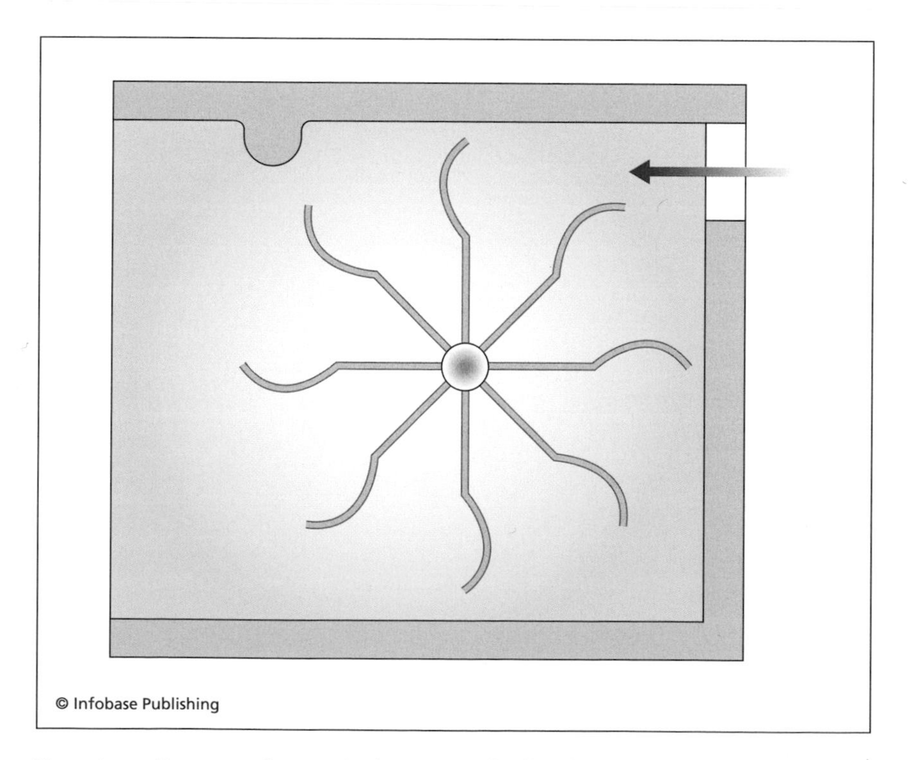

Top-view diagram of a vertical axis windmill—the wind blew through the opening on the right and blew out through the opening on the left

walls: As the wind blew on the sails on one side of the *windshaft* (the axle to which the sails are affixed), it would also blow on the sails on the other side of the windshaft. The result would be that the force on one set of sails would be balanced by the force on the other side and little or nothing would happen. The solution adopted by the Persians was to enclose the windmill within a walled structure that had one opening through which the wind could enter and one opening, located on the opposite side of the structure, through which the wind could exit. The entrance was sometimes conical in shape so that as the wind blew into the wide end of the hole, it accelerated, and emerging with increased velocity from the narrow end, it pushed against the sails. The opening was placed so that the wind was directed toward one side of the windshaft only. The sails on the other side of the shaft were protected from the wind by a curved or L-shaped wall. This allowed the wind to push preferentially on one side of the windshaft and so cause it to turn in one direction only.

The main advantage of this design is its simplicity. There were no gears. By mounting the sails on a vertical shaft, the first windmill builders could attach millstones directly to the spinning shaft. The wall might seem to be a significant disadvantage because if the wind changes direction it will not pass through the openings, and consequently, no power will be generated. These early mills were, however, built within an area of Persia that today is called Sistān and is located near the Iran-Afghanistan border. In Sistān, the wind blows steadily from the north from mid-June until mid-October. Provided the mills were built facing north, they did not have to be sensitive to changes in the wind because there were none. One famous site in the ancient Persian city of Neh had 75 of these vertical-axis windmills, all built in a single long row.

The main disadvantage to the vertical-axis design lies in the fact that it makes inefficient use of the sails. Because the wind only blew on half the sails at one time, only half the sails were transmitting force to the spinning shaft. The other half was, at best, dead-weight. Moreover, even on the side of the shaft on which the wind blew, some sails partially obstructed others from the full force of the wind. Vertical-shaft windmills did not exert much force; their output, when measured by the amount of processed grain they produced per hour, was not large.

Despite its limited efficiency, the vertical-shaft windmill has never fallen entirely out of favor. Traditional designs were used in parts of Iran and Afghanistan well into the 20th century, and Europeans and Americans have experimented with the design for centuries. The British physician Erasmus Darwin (1731–1802), grandfather to the naturalist Charles Darwin, built a design of which he was very proud: It consisted of a cylindrical tower equipped with slats that could open and close. The (vertical) windshaft was placed at the center of the tower and spanned its entire length. The sails, which looked more like ceiling fans than revolving doors, were attached to the top of the shaft. The slats on the windward side of the tower would be opened at an upward angle. The wind would, therefore, pass through the slats, and as it passed through the slats, it was deflected upward through the tower. As the wind exited the top of the tower, it acted against the sails and set the shaft turning. Passing the wind through narrow openings and changing its direction from horizontal to vertical, not surprisingly, diminished the force of the wind. This design was, therefore, capable of only modest power even in a strong wind. (A more recent vertical-axis design, called a Darrius wind turbine, looks like an eggbeater and may prove better suited for use in tidal mills.)

The history of traditional horizontal-axis European windmills begins during the 12th century. The origin of the European design is unknown. Because the first European references to windmills occur after the First Crusade, which began in 1095 C.E., some think that the concept, at least, is of Middle Eastern origin, but even the earliest European designs employed an almost horizontal windshaft, a design quite different from the vertical windshaft designs

Post mill. Because the entire structure was mounted on a single pole, it could be turned to face the wind no matter from what direction the wind blew.

that were still in use in the Middle East. European mills attempted to harness the wind regardless of the direction in which it blew, a very different goal from the early Persian design.

The earliest type of European windmill is called the *post mill.* It was an extremely durable technology. A few commercial post mills—they were used to grind grain—were operated in Great Britain well into the 20th century. Post mills were, therefore, economically viable for about 800 years. These mills were, for the most part, used to grind grain or pump water. The main structure of a post mill, the enclosure in which the miller worked and which housed the millstones, looks like a shed or small house supported above the ground by a single large round post. The windmill's sails were mounted on a shaft that was inclined slightly upward—often at about 15°. The windshaft was coupled by two or more gears to a vertical shaft to which the millstones were attached.

Compared to Persian windmills, the post mill had the potential to be more efficient because it used all of its sails all of the time—or at least all of the time that the wind was blowing. Early sails were made of large wooden lattices, each of which was attached to the windshaft by a single spar and covered with canvas to provide a surface against which the wind could push. (Etchings show that very early windmill operators actually wove pieces of canvas through the lattices.) The sails were slightly tilted relative to the plane in which they rotated so that when the wind blew on the sails, it imparted a turning force to the windshaft. (To see why this works, imagine that the sails were oriented so that they lay in their plane of rotation, and suppose that the windshaft pointed directly into the wind. In this configuration, the entire force of the wind would be caught by the sails and transmitted directly back along the windshaft without exerting a turning force.)

The output of the windmill was sensitive to the direction of the wind. A shift in wind direction could leave the canvas flapping ineffectually on a stationary shaft. This is the reason that the post mill was mounted on a post. The entire structure could pivot on its post as if it were a multiton weathervane. The turning procedure was originally accomplished manually: A long pole extended out from behind the mill. The pole was simply a very large lever. When the direction of the wind changed, the mill operator went behind the mill and pushed on the far end of the pole until the windmill faced into the wind again. Later a secondary sail was sometimes added to the back of the mill. It was oriented so that it was parallel to the windshaft. Its purpose was to provide a surface against which the wind could push so that the mill would pivot without operator intervention.

As one might expect, the larger the area of the sails, the more of the wind's power could be transmitted to the windshaft, and the more work the mill machinery could perform per unit time. All other things being equal, therefore, larger sails were better. But larger sails required a larger support structure, and, in particular, the structure had to be taller because the axis on which these longer sails turned had to be lifted higher off the ground so that the spinning sails could clear the ground as they rotated. But this was only one factor that caused designers to build taller mills. At least as important is the fact that winds along the ground are weaker than those at higher elevations due to interaction of the wind with trees and buildings, a fact long known to windmill designers. This resulted in two trends in windmill design: Mills became more massive so that they could resist the force of the wind against the larger sails, and they became taller in order to raise the sails as high into the oncoming wind as possible.

Building a very large post mill is problematic. As the structure becomes more massive, it becomes more difficult to support on a single post as well as more difficult to turn. The alternative to the post mill was the *tower mill.* (The so-called smock mill is usually listed as a different type of windmill because from the outside it looks different from the tower mill—its shape is supposed to be reminiscent of a smock—but mechanically, tower and smock mills operate according to the same general design. This book will discuss only the older tower mills, but most of the statements made about tower mills apply word for word to most smock mills.) The tower mill was developed during the late Middle Ages. They were anchored securely to the ground and were often several stories tall. Only the top section of the mill, called the cap, turned so that the sails could face the wind. The cap housed the windshaft. The much larger bottom section, which was often large enough for storage and living quarters, was immobile.

Tower mills were often taller and more massive than post mills. The biggest were approximately 100 feet (30 m) tall. At this scale even the cap was massive, and the track on which the cap turned had to support the massive sails and the windshaft, in addition to the cap itself. To reduce stress on the track and to make the problem of turning the cap manageable, tower mills were usually built in the form of truncated cones—that is, the round base is massive and wide, and the sides slope upward to a smaller and relatively lighter cap. As with their smaller counterparts, tower mills continued to make an important contribution to the economies of several European nations for several centuries and only began to fade late in the 19th century.

Throughout the history of European windmill design, the main technical issue, for economic and safety reasons, concerned the design of the sails. The millwright had to be able to control the amount of the wind's force that was transmitted to the sails. An unexpectedly powerful wind could cause the sails to turn too fast. The flour, the processing of which was the main reason that many of these windmills were constructed, could be burned if it was milled when the machinery was spinning too rapidly. Moreover, a rapidly turning windshaft could damage or destroy equipment. And working in a mill experiencing unexpectedly strong winds was dangerous too. Sometimes, the millwright was injured or even killed as unexpectedly high winds spun the mill's machinery out of control.

There were also the problems associated with working in light winds. For economic reasons the mill should operate whenever the wind was blowing. But in order to overcome static friction, stronger forces were necessary to start the windshaft turning than were needed to continue operation once the sails had begun to rotate. Poorly designed sails could prevent the windmill from starting in light winds.

Dutch tower windmill (no longer functional) *(Harskampermolen Mariana)*

To control the amount of the wind's force transmitted to the mill, millwrights developed methods for adding or reducing the amount of canvas exposed to the wind. Recall that the canvas covered a lattice structure, so that using canvas to cover only part of the lattice enabled the millwright to continue to operate the mill at a safe speed under heavy winds. When the winds were lighter, more canvas was deployed. Several ingenious methods were developed that enabled the millwright to adjust the amount of canvas without having to climb all over the sails.

Changing the amount of sail exposed to the wind by altering the amount of canvas was still time-consuming and occasionally dangerous. Gusts of wind could still accelerate the sails in unexpected ways and occur too rapidly for the operator to respond. A later invention involved replacing the canvas with devices that looked like venetian blinds mounted on springs. The tension of the springs could be adjusted from inside the mill via levers. When the wind blew harder than some predetermined level of force, it blew the blinds open, and the wind "spilled" through the lattice without generating any turning force. When the force of the wind was below that predetermined level, the springs closed the blinds, providing a surface against which the wind could push.

With respect to the problem of starting the mill in light breezes, the best way was to design sails that were strongly tilted away from the plane of rotation. This enabled the wind to exert a strong sideways force—strong enough for a light breeze to begin turning the sails. The problem is that as the speed of rotation increased the tilted sails actually slowed the spinning shaft, decreasing its efficiency. The problem was most pronounced at the ends of the sails, which, after all, always turn faster than those parts closer to the windshaft. The solution was to twist the surface of each sail so that near the windshaft the sail's surface was turned strongly away from the plane in which the sails rotated. Nearer the tip, however, the surface of the sails twisted so that they were more nearly parallel to the plane of rotation. Designers believed that this design enabled the mill to start turning in a light breeze, and still operate efficiently as the speed of rotation increased. Various forms of several of these innovations can still be found on contemporary wind turbines.

Until late in the 19th century, tens of thousands of windmills dotted the landscape of many parts of northern Europe. In 1900, Finland, for example, had 20,000 operating windmills. Large numbers of relatively low-power windmills distributed over a large area was an almost ideal solution to the demands of the time. Throughout most of the history of the European windmill, the transportation infrastructure of most European nations was not especially good. Grain was, therefore, grown and milled locally, and, provided a windy site was located nearby, windmills were a good match for this type of small-scale production.

The last large-scale effort to convert wind power into mechanical energy—as opposed to electrical energy—occurred in the United States. Farmers, faced with the vast treeless expanses of the Great Plains, a region that often lacked sufficient surface water for their agricultural needs, installed windmills to pump water from the ground. These windmills used so-called annular sails, numerous curved wood or steel slats that radiated out from the center. The sail was held perpendicular to the wind by a vane in back. The tower often consisted of an unadorned steel lattice that supported the sail at a high enough height for it to operate efficiently. During the latter days of the 19th century, hundreds of American firms produced hundreds of thousands of these wind-driven pumps. These devices are still manufactured. Currently there are an estimated 1 million to 2 million wind-driven water pumps in service worldwide, but their value to agriculture has been surpassed by pumps powered by electricity, gasoline, or diesel fuel, all of which produce more power and operate without regard to the weather.

WIND TURBINES

The term *wind turbine* is usually reserved for windmill-like machines that convert the kinetic energy of wind into electricity. Perhaps the first such energy-conversion device, designed and built by the American engineer and inventor Charles F. Brush (1849–1929), is known as the Brush turbine. It was erected in Cleveland, Ohio,

Lely wind farm, the Netherlands, one of the first offshore wind farms *(United Nations Atlas of the Oceans)*

and began operation in 1888. The sail was 50 feet (19 m) across and provided electricity to Mr. Brush's very large house. Brush solved the problem of windless days by using the modest output of his enormous machine to charge batteries. His invention was highly acclaimed—there is a detailed description of the Brush turbine in the December 20, 1890, issue of *Scientific American*—but the idea was not widely copied. Even Brush turned his attention to other projects.

The earliest sustained effort to harness the wind was undertaken in Denmark beginning in 1891 by Poul la Cour (1846–1908), a prominent Danish inventor and teacher. La Cour's efforts were carried out over several decades; his work was supported by the Danish government, and his discoveries had a more lasting effect than those of Brush. It is interesting to note that la Cour's first solution to the problem of power production on windless days was to use the electricity produced by his wind turbine for electrolysis—the separation of water molecules into oxygen gas and hydrogen gas. The hydrogen could be stored and burned at the convenience of the user, and the resulting heat could be harnessed as desired.

Modern grid-connected wind turbines have a great deal in common with the windmills that preceded them. The rotor, the analogue of the sail, is usually mounted on a horizontal shaft. The shaft is often connected to a transmission, called a gearbox. The gearbox is placed

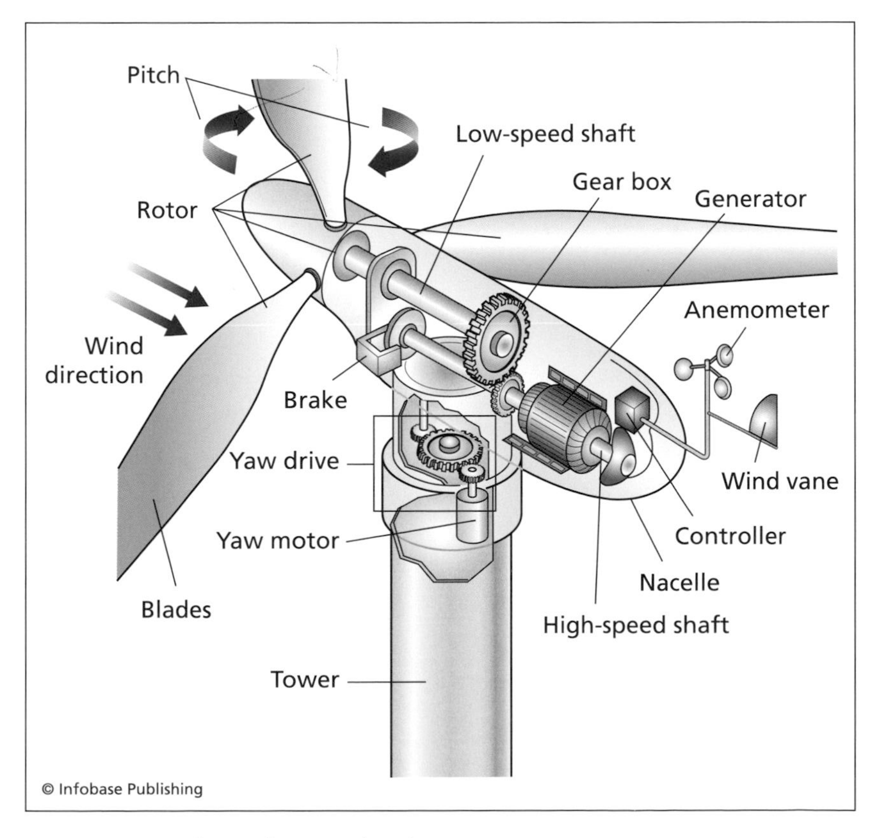

Cutaway view of a modern wind turbine

between the windshaft and the generator and is used to ensure that the generator turns at an optimal speed. From a distance, the blades of a wind turbine may appear to turn fairly quickly, but in order to produce electricity suitable for the electric grid, the generator must turn much faster—typically in the 1,200–1,800 rpm range. This change in speeds is accomplished by the gearbox. (Early windmills sometimes used sets of gears to change the rate at which the millstones turned, but the difference in rotation rates was typically much less than that obtained with modern wind turbine gearboxes.)

Just as larger sails were preferred to smaller ones by designers of traditional windmills, contemporary engineers prefer longer rotors to shorter ones and for the same reason: The amount of power converted is proportional to the area swept out by the rotor. In order to capture as much of the wind's energy as possible, the wind turbine should use the longest rotor practicable.

Height is another goal of wind turbine engineers just as it was of their predecessors. Recall that early windmill designers built tall windmills in order to raise the sails as high into the wind stream as possible. Contemporary wind turbine designers also mount their turbines on towers that are as high as practicable. Today, engineers often mount turbines on towers that exceed 275 feet (84 m) in height. Is the extra height worth the expense? A rule of thumb is that on an open unobstructed landscape, wind speed will increase by about 25 percent as elevation above the ground increases from 50 feet (15 m) to 200 feet (61 m).

Another area of commonality involves the problem of regulating rotor speed. Wind currents are highly variable, and just as windmills were designed to operate within certain limits, so, too, are wind turbines. A very strong wind has the potential to spin the rotor too quickly, which could result in expensive repairs. The rotors on wind turbines are aerodynamically designed to produce a pressure difference between one side of the rotor and the other as wind flows past the rotor. The idea is similar to that used to create lift with an airplane wing. It is this pressure difference between the two sides that causes the rotor to rotate. When the wind becomes too strong, a control device changes the angle of the rotor relative to the wind causing the pressure difference to diminish. In this way the forces acting on the turbine are controlled, and the wind turbine is protected from wind damage.

An early version of this protective technology was a spring-activated linkage that was similar in concept to the spring-mounted venetian blind–like devices on traditional windmills described earlier. When the rotor began spinning too quickly the spring linkage was released, and the angle at which the rotor met the oncoming wind changed. The lift forces were eliminated, and the wind turbine was made safe. This task is now usually handled by computer, which monitors the rotor and continuously adjusts the angle at which the rotor meets the oncoming wind in order to keep the forces acting on the rotor within prespecified limits.

One of the main differences between traditional windmills and contemporary wind turbines involves the speed with which they rotate. The simple construction materials used in early windmills meant that the windshaft could not rotate quickly. But a slow rate of rotation—slow, that is, relative to the velocity of the oncoming wind—also meant that the windshaft exerted a powerful turning force. This is exactly the characteristic that is desirable for applications such as pumping water and processing grain. High rates of rotation are better suited for electricity generation. The speed of the tip of a modern three-rotor wind turbine will typically turn between four times and six times the speed of the wind that drives it. A very long rotor need not turn at a high rpm to achieve its optimal ratio between tip speed and wind velocity. A shorter rotor, by contrast, must turn much faster in order to maintain that ratio. This explains why smaller wind turbines typically rotate faster than larger ones.

Achieving the goal of a high rate of rotation in moderate wind speeds also helps to explain why wind turbine blades look so much different from the sails one finds on traditional European windmills or the sails one finds on the windmills that once were so common on the Great Plains of the United States. Obtaining a high rate of

The Altamont Pass Wind Resource Area

Altamont Pass, located in California, is one of the earliest and at one time was one of the largest of all wind farms located in the United States. Altamont Pass was chosen because it is located near a high-voltage transmission corridor—building a high voltage line to connect a power production facility to the grid is expensive—and because high temperatures inland often result in the formation of a wind that blows-from the ocean through Altamont Pass. Just as important—perhaps more important at the time that Altamont was constructed—the state of California offered substantial tax incentives to companies that built wind-power facilities in the state. Altamont Pass was once widely hailed as an environmentally friendly producer of electricity.

Since 1981, when construction began, thousands of wind turbines have been built here. They are built along ridges and other locations where the probability of strong winds is optimized. The sight of these wind turbines was so impressive that Altamont Pass became an important tourist destination. Row after row of moderately sized turbines spinning furiously in the wind was, at one time, a sight that many found fascinating.

By today's standards, many of the turbines at Altamont Pass are small and inefficient, and they are being replaced with a smaller number of larger, more efficient turbines. But even with old technology, in 2005 Altamont Pass had a total rated capacity of 548.3 MW—that is, if all the turbines were operating at full power, Altamont Pass would generate 548.3 MW of power. Often, however, not all of the wind turbines operate at full power, and there are some days when none of the turbines operate at full power because there is not enough wind.

How valuable a project is Altamont Pass? For planning purposes, the Monterey Bay Regional Energy Plan estimates that the percentage of time that wind power resources produce their rated power is 20 percent. This

rotation requires a small number of long thin blades. Obtaining a slow motion with a high turning force is best accomplished with wider, more numerous sails. While modern wind turbines typically

includes the Altamont Pass project. There is a big difference between the so-called *nameplate capacity,* which is what a project generates when it operates at full power, and the actual power output, which is always much smaller.

The biggest criticism of Altamont Pass concerns its environmental impact. To be sure, when the wind is blowing, the turbines produce emissions-free power, but they also regularly kill golden eagles, burrowing owls, and red-tailed hawks. These birds, and others, fly into the rotors. The issue of the destruction of bird species has become one of the central issues associated with the operation of Altamont Pass.

Studies have shown that bird mortality can be reduced by changing the design of the towers. Some tower designs attract more birds than other designs, causing the birds to perch near the potentially lethal rotors. And studies have shown that sometimes the location of a wind turbine makes it particularly deadly: Turbines located in canyons tended to destroy more golden eagles than other similar turbines placed elsewhere. Presumably, mortality can be reduced by building replacement units with less attractive towers, and excluding some locations from development. But this may not be enough to satisfy many opponents because the wind farm is also located along a bird migration route, and some deaths of the golden eagle, a federally protected species, seem unavoidable.

Some groups now advocate shutting down Altamont Pass during the winter months, when bird mortality is highest. This proposal, if enacted, would further diminish the economic value of the wind farm. Altamont Pass illustrates the difficulties of identifying what constitutes "green energy." Perceptions of what it means to produce power in environmentally sound ways continue to evolve.

use two or three thin blades, traditional windmills use four or more wide sails.

Large modern wind turbines can, under the right conditions, generate significant amounts of power. Details vary by manufacturer, but General Electric, for example, currently manufactures wind turbines in the 1.5–3.6 MW range. When all its occupants are home and all of its appliances are drawing power, the average home will draw anywhere from two to four kilowatts. This is the maximum rate at which it draws power. An electrical grid must always be prepared to meet maximum demand. (A system that only meets the average demand will fail to meet the demand for power roughly half of the time. Most people would agree that a 50 percent failure rate for an essential service is unacceptable.) A three-MW turbine will, therefore, be able to provide enough power for between 750 and 1,500 homes, provided the wind is blowing. Clearly, powering a modern national economy with wind turbines would require an enormous number of wind turbines even when the wind is blowing hard, and when the wind is not blowing, no collection of wind turbines is enough. How, then, can one assess the value of wind turbines? How much reliance can be placed on a power source that is inherently unreliable? These questions are addressed in the next chapter.

The Nature of Wind Power

Wind turbines are energy conversion devices. They convert the kinetic energy of the wind into electrical energy. This simple-sounding statement indicates that there is an upper limit on the amount of energy that can be derived from the movement of a particular mass of air. No wind turbine can produce more electrical power than the amount of power in the wind itself. In fact, the amount of electricity produced by any turbine must be substantially less than the kinetic energy of the wind, since any device that converted all of the kinetic energy of the wind into electrical energy would have to stop the wind from blowing. Determining the maximum amount of energy that can be produced per unit time from moving air—that is, determining an upper limit on the power generated by a wind turbine—is an important first step in understanding the potential contribution that wind turbines can make toward the electricity supply. This is one goal of this chapter.

Because winds at higher altitudes are stronger and steadier, the towers are as tall as is economically practical. *(Cam Gordon)*

But there is more to understanding wind energy than computing the amount of energy available on a windy day. Some days the wind does not blow. Wind turbines are, therefore, fundamentally different from coal and nuclear plants in that wind energy is intermittent by nature. While it is true that coal and nuclear plants may break down without warning, experience has shown that both types of power plants produce power reliably and continuously for long periods of time without interruption. Consequently, they are usually used to meet the minimum demands of an electricity market. This is called base load power. Wind energy is poorly suited for this type of function. Wind does not blow steadily enough—wind is not reliable enough—to supply base load power. Consequently, wind does not compete with coal and nuclear.

There is, however, another aspect of electricity production called peak power. (Power is sometimes further subdivided, but for purposes of this discussion, as for many others, it is sufficient to divide power production into peak and base load.) Peak power is the electricity that must be produced to supplement the base load requirements on an hour-by-hour basis. Peak demand occurs daily as electricity use rises during business hours or on hot days. (Air conditioning, for example, is an energy-intensive technology.) Peak power requirements usually occur for just part of each day. Natural gas plants are often used to meet peak demand, and, in fact, some natural gas–fired power plants are used only during peak periods. The reason, as previously mentioned, is that natural gas has become so expensive that electricity produced by natural gas–fired plants is, when compared to that produced by coal and nuclear plants, uneconomical to produce. Each day, as demand rises, the order will go out to turn on natural gas–fired plants so that they will be ready to produce power at the required time. If the demand is high enough, natural gas plants will be brought online to produce electricity for the grid, and the owners will be paid accordingly. Later in the day, as demand falls, the plants are turned off. In contrast to natural gas plants, wind turbines produce power (or not) independently of the demand for electricity because the wind blows without regard for the electricity markets. In this sense, wind turbines are not well-suited to peak power production either, because (again) they are unreliable. And yet wind turbines do produce power, and sometimes many wind turbines produce a great deal of power. This raises the question of how one should value the contribution that wind power can make. Answering this question is the second goal of this chapter.

Finally, this chapter discusses the environmental effects of wind power. Every technology that produces commercially significant amounts of electricity has significant environmental effects. Wind power is no exception.

HOW MUCH ENERGY IS IN THE WIND?

Ultimately, the winds are driven by the unequal heating of Earth's atmosphere. Wind is, therefore, a form of solar energy. Earth spins rapidly—land at the equator is moving about Earth's axis of rotation at approximately 1,000 miles per hour (1,600 kph)—and because Earth is a sphere, different latitudes are exposed to differing amounts of sunlight for different periods of time. And different parts of Earth's surface absorb the Sun's heat in different ways. The oceans, for example, absorb solar energy differently than land does. Finally, air is transparent to the Sun's light in the sense that as the rays from the Sun pass through Earth's atmosphere, they have little effect on its temperature. Instead, the Sun's rays heat Earth's surface, which absorbs some of the Sun's light and radiates it back into the atmosphere as heat. (Some light is also reflected.) Earth's atmosphere is, therefore, heated from below, not from above. The uneven heating causes changes in the density and pressure of the air near Earth's surface, causing the air to move. The path that large-scale motions of air take across Earth's surface are influenced by the planet's rotation. This very dynamic process produces moving regions of low- and high-pressure air that form and dissipate and in the process produce winds. Some winds are local, and some sweep over large areas of Earth's surface. The processes involved are as old as the planet on which they occur.

As with hydropower, wave power, and tidal power sites, windy sites are natural resources, the locations of which are beyond human control. Simply put: Windy sites can be developed or not, but they cannot be created. The suitability of each site should, of course, be evaluated on its own merits, but the number of commercially viable sites is not as large as one might think. The decision to forgo development of one windy site for environmental, economic, or aesthetic reasons is also a decision to either attempt to extract more power from another site, assuming one is available, or to switch to another technology to produce the electricity that no one is willing to live without. (The problems associated with site development are considered in more detail in the next chapter.)

Winds are a three-dimensional phenomenon, and the characteristics of wind depend very much on the elevation above the ground at which it is observed. Winds that are close to the ground are very

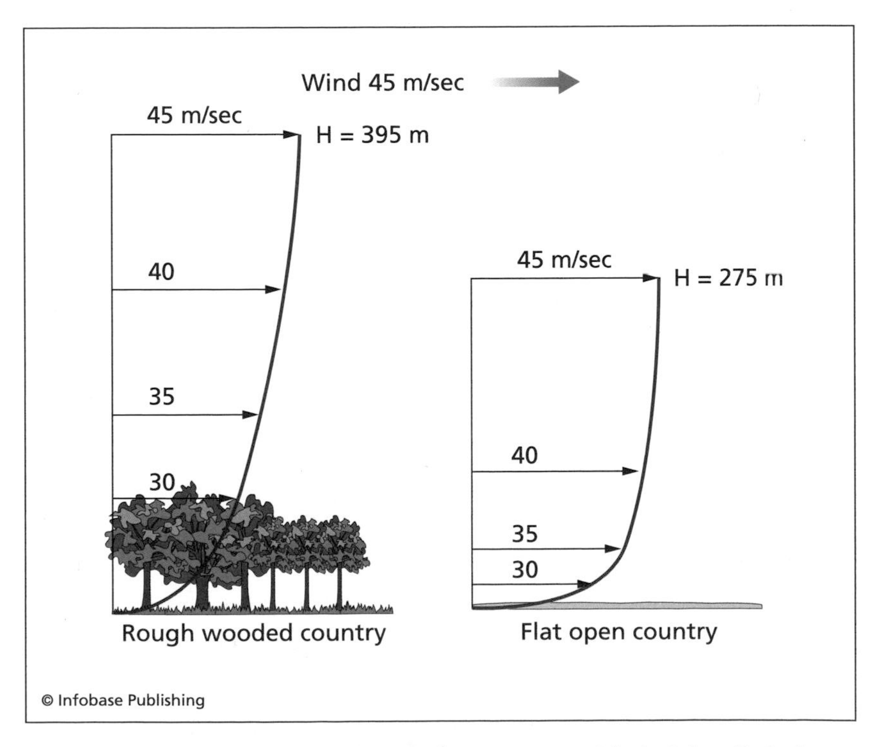

Diagram showing how wind speed increases with height. Turbulence decreases with height.

turbulent. They are accelerated as they pass between tall buildings or through mountain passes. They are deflected by buildings and trees, and as the wind passes over and around these obstacles eddies form and roil along at ground level. This is most easily observed during blizzards when snowflakes are swept along the complex, uncertain paths that the wind takes as it moves across the landscape.

Turbulent air movements contain a great deal of kinetic energy, but because much of this energy is contained in eddies and other small-scale unsteady motions, it is difficult to convert the kinetic energy of turbulent airflows into electrical energy. Moreover, as the wind blows through trees and across "rough" terrain, some of the energy of the wind is dissipated. It is a much easier process to convert into electricity the smooth steady flows of air that occur hundreds of feet above the ground. At higher altitudes the wind is unobstructed and less turbulent, and so the process of converting its kinetic energy into electricity is more efficient. This is one reason that commercial wind turbines are so tall. They tower over the landscape in order to lift the blades of the turbine into the smoother, steadier airflows that exist hundreds of feet above the ground. Turbines mounted on shorter towers are less productive, less efficient, and less profitable, and consequently when mounted on shorter towers more turbines are required to produce the same energy as fewer turbines mounted on taller towers. There is, as a general rule, little economic sense in mounting an expensive turbine on a short tower.

Companies seeking to produce electricity from the wind must identify sites where the wind blows as steadily as possible. If they receive approval to erect their turbines at such a site, they construct towers that extend high enough to capture wind unperturbed by interaction with the ground. Under these idealized conditions, how much power can a wind turbine of a given size generate when wind is blowing at a given speed?

The answer to the preceding question is contained in equation (5.1), the same equation that was used to determine the maximum amount of power that can be harnessed from ocean currents. That equation is repeated here for ease of reference:

$$P_{max} = \frac{CdAv^3}{2} \quad \textbf{(9.1)}$$

In this equation, P_{max} still represents the maximum amount of power that can be obtained from a turbine, but now d is the density of air, A is the area swept out by the spinning (wind) turbine blades, v is the speed of the wind, and C is a number that represents the maximum percentage of the wind's energy that can be harnessed. (The expression $dAv^3/2$ is the total amount of power contained in air of density d blowing through a tube of cross-sectional area A with velocity v.) Equation (9.1) reveals a great deal about the physics of wind turbines.

First, d, the density of air, is very small compared to, for example, the density of seawater. To see the difference between air and water—and so the difference between wind turbines and tidal mills—it is enough to know that the density of seawater is roughly 900 times greater than the density of air at sea level. This is one reason why tidal mills can, in principle, produce so much electricity from slowly moving ocean currents: Although the velocity of the ocean current is usually small compared to the velocity of wind currents, the amount of power that can be generated by an ocean current can still be large because of the relatively high density of seawater.

Second, the density of air decreases rapidly with height so, for example, although the wind may blow steadily at the top of a tall mountain, there will not be as much energy to convert at that altitude as there is at lower altitudes—all other things being equal—because of the thinner mountain air. To take an extreme case, the density of air at the top of Mount Everest is only about one-third the density of air at sea level.

Third, equation (9.1) shows that the amount of power that can be harnessed from the wind is highly sensitive to the velocity of the wind. As noted in chapter 5, if the speed, v, of the wind is doubled, the power of the wind increases by a factor of 8 ($2v \times 2v \times 2v = 8v^3$). To take a more realistic example, a wind turbine operating in a wind that blows at 12.5 miles per hour (20 k/hr) can, in principle, generate twice the power of a turbine operating in a 10-mph (16 k/hr) wind. (As will be shown later, wind turbines are designed to produce a rated power output, so the situation is more complex than that shown here. Equation [9.1] simply indicates that no matter how one designs a wind turbine, the amount of power the turbine produces cannot exceed the amount predicted by this equation.)

Fourth, assuming that a power company can find a suitably windy site that is reasonably close to its target market, it will want to make sure that the turbine blades sweep out the largest area possible. This is done by equipping the turbines with the longest possible blades. The area, A, that is swept out by the rotor is a circle that depends on r, the distance from the tip of the blade to the hub, according to the following formula—a formula that every student encounters in geometry:

$$A = \pi r^2$$

where π is a number that is a little larger than 3. By doubling the length of a blade—that is, by doubling r—this formula shows that the area swept out by the rotor will quadruple ($2r \times 2r = 4r^2$). Consequently, doubling the length of a blade quadruples the available power. This explains why it is to the designer's advantage to make the blades as long as possible. The need for longer blades further drives up the height of the tower. In order for the turbine to achieve maximum efficiency, the lowest point reached by the blade should be as high above the ground as practicable.

This leaves only the question of the value of C, and the answer is that C is the same regardless of whether one considers wind currents or water currents. As mentioned in chapter 5, during theoretical investigations conducted at Göttingen University in Germany during the 1920s, the German engineer and scientist Albert Betz estimated that the symbol C in equation (9.1) was no larger than 0.59, which means that no more than 59 percent of the kinetic energy of the wind can be converted into electricity, and contemporary research indicates that C is probably quite a bit less than 0.59. Estimates vary.

Power producers that depend upon wind are especially concerned with finding places where the wind blows reliably and rapidly. Their choices are further constrained by the necessity of building on sites where they will have access to high-voltage power lines. (Building high-voltage lines is very expensive and always controversial. Success in obtaining the necessary permits is not assured.)

Assuming, then, that a power producer receives approval to build a wind farm and (if necessary) the high-voltage power lines necessary to connect its site to its target market, the next question is how should the utilities that might buy power from the power producer compute the increase to system reliability that the wind farm offers? A utility must produce power simultaneously with demand. The reliability of the grid is important from a planning, a safety, and a legal standpoint. In many areas there are legal requirements that utilities maintain a reserve capacity at all times in order to meet unanticipated events—either a surge in demand or the failure of a particular power source. System reliability is extremely important. How do wind farms contribute to system reliability?

Although it is true that no power source operates indefinitely without interruption, conventional power sources—especially coal, nuclear, and natural gas plants, the main sources of electricity in the United States and most other nations—are much more reliable than is the wind. When a conventional plant is turned off, it is usually because of regularly scheduled maintenance. Predictability is one of the key advantages of conventional power plants. To make use of the wind, it is necessary to find a way to integrate an inherently unreliable power source into a system that must supply power with a high degree of reliability.

ESTIMATING CAPACITY

There is not yet a definitive answer to the problem of determining what sort of contribution to the reliability of an electric grid a wind farm makes. Different utilities and system operators place different values on wind power. Differences in opinions arise because wind power is often unavailable exactly when it is most needed—and here the word "often" is relative to more conventional power producers such as natural gas plants. But while it is evident that wind is not as reliable as more conventional power sources, it has some value, because it *sometimes* produces electricity when electricity is most needed. This assertion does not, of course, identify the value that one can assign to wind power; it only claims that wind power has more than no value. But those who operate power distribution systems require a more precise statement in order to ensure the reliability of the network. Reliability is one of an electrical network's most important properties.

An important characteristic with respect to wind turbine reliability is its capacity factor. There are multiple definitions of capacity factor, but a common one is defined to be the actual power output of the turbine divided by the maximum possible turbine output over the course of a year. The capacity factor is, then, the percentage of the maximum output actually produced by the turbine in question during one year of operation. The advantage of such a measure is its simplicity. Its disadvantage is that it fails to entirely capture the practical contribution made by a turbine.

In the business of power production, timing is everything. Because electrical power cannot be stored for later use, it must be produced simultaneously with demand. A power source that cannot produce power when there is a demand for it is, it is generally agreed, worthless. Very sophisticated systems are now in use to

Large wind farms require large amounts of land. *(Indiana Office of Energy and Defense Development)*

track changes in demand and to use the least expensive supplies to satisfy those demands. Provided that a wind farm is producing power at a time when there is demand for it, that power can be used in one of two ways: Either it can be used to meet demand, or, alternatively, the wind farm can—for as long as the wind is blowing—be held in reserve to meet *unexpected* demand. Reserve power is also important. All electrical power networks maintain a reserve capacity, a safety margin, to ensure that demand will be satisfied in the event that it were to suddenly increase or in the event that online supplies were to suddenly decrease, a situation that would arise if, for example, there was an unscheduled shutdown of a power producer. In many states, the obligation to maintain sufficient reserve capacity, which may be as high as 15 to 20 percent of peak demand, is a legal requirement. Electrical networks that operate without a reasonable reserve capacity do so at their peril. California, for example, had 2 percent reserve capacity during the year 2000. During this time its network was highly unstable. The unpredictability of the network was reflected in power blackouts and sharp increases in electricity prices. Since that time, a great deal of attention has been paid to creating and maintaining adequate reserve capacity for the California system.

Some analysts say that a power source should be valued in accordance with its contribution to system reliability. A power source, no matter whether it be wind, water, natural gas, coal, nuclear, or "other," only has value if it contributes to system reliability. The more reliable the power source is, the more valuable it is. Unlike the more simplistic definition of capacity described earlier, the approach about to be described takes into account both the amount of power produced as well as the time at which it is produced.

As described in chapter 2, electrical power is usually divided into base load power and peak power. Every electrical market has a minimum power requirement that is characteristic of that market. This minimum requirement is called base load power. Late at night, for example, or early in the morning, base load power is all the power that is needed to meet the demands of the marketplace. But additional power—that is, power that exceeds base load demand—is usually required during the day as schools, banks, stores, and other predominantly daytime consumers of electrical power begin to power up. Electricity used during periods of higher demand is called peak power.

Utilities generally sign long-term contracts for base load power because the demand for base load power is highly predictable. Peak load power is less predictable, and it is often purchased day-to-day or even hour-to-hour on what is called the spot market, and it is generally supplied by other types of power-producing technologies than coal or nuclear. Wind power, because of its intermittent nature, is poorly suited for base load power, but it may be useful for peak power production.

It is important to emphasize that no power source is perfectly reliable. Each power source is occasionally unavailable because of scheduled maintenance or because of mechanical failure. Mechanically speaking, wind turbines are very reliable. They are, on average, mechanically available more than 95 percent of the time, which is another way of saying that they are broken or shut down for maintenance less than 5 percent of the time. But wind farms have an additional source of uncertainty. Wind, which is their "fuel," is only intermittently available. But by itself this observation tends to overstate their unpredictability. Although it is not possible to predict wind speed one or two months into the future, weather reports can be useful in predicting wind speeds one or two days into the future, and such reports are especially useful for predicting wind speeds in one or two hours. Accurate weather forecasting can enable wind producers to participate in the spot market with some confidence. (The spot market is the day-to-day and hour-to-hour market that exists to meet peak power demands.) With accurate forecasting and good timing, wind producers can bid for contracts and earn profits in the spot market. The amount of profit depends on the timing of the wind. Given two wind farms, if the wind usually blows during peak hours at one site and during off-peak hours at another, the site that produces power during peak hours is more valuable than the off-peak site *even if the off-peak site produces somewhat more total power.*

Wind data is the first requirement for estimating the potential value of a wind turbine site. Measurements are taken hourly for at least a year. This enables engineers to estimate both the amount of power available at the site and the times that it will probably be produced. At most sites, some times of the year, and even some times of the day, are reliably windier than others. Using this data, engineers will estimate the amount of power that the wind turbine will probably produce and when it will probably produce it. They compare that estimate with a more predictable source of power called the benchmark. The benchmark, which is a mathematical idealization of an actual power plant, is often taken to be a natural gas plant because natural gas plants are also often used for peak power production and their performance characteristics are well understood.

Based on wind measurements and the known mechanical reliability of wind turbines, engineers compute the contribution to system reliability that the wind turbine will probably make. Their answer depends on how much power will probably be obtained from the wind at the site as well as the times that the wind will probably be blowing. (These calculations require probability theory, that branch of mathematics that seeks to quantify uncertainty.) Next the power output from the benchmark producer is adjusted until its contribution to system reliability equals that of the wind. The contribution is calculated based on the idea that a more powerful but intermittent source of electricity (wind energy) makes the same contribution to system reliability that is made by a lower power, more reliable source (the idealized natural gas plant). The calculated output of the benchmark is noted, and becomes the *effective load-carrying capability* of the wind farm. This is also sometimes called the capacity of the wind farm. (The word *capacity* is used in several different ways.) This capacity rating is the computed value of the wind farm to grid reliability.

Depending on the site, wind turbines are generally valued between 20 and 40 percent of their full power output. Turbine manufacturers and turbine enthusiasts often quote a higher number, the maximum power output of a turbine, also called the nameplate capacity, as evidence of a turbine's value as an emissions-free generator of electricity. But system planners and operators are guided more by the effective load-carrying capability (or some other conceptually similar measure) because it is a more accurate measure of the contribution that can be expected from the turbines in actual practice.

To see the difference between the two estimates, it is enough to note that some wind enthusiasts will assert that wind turbines can be substituted for conventional sources on a megawatt-for-megawatt basis. By way of example, one might support replacing a 100-MW natural gas-fired plant with 100 individual one-MW wind turbines, but the preceding discussion shows that this is almost certainly not a reasonable suggestion. A more realistic estimate is that it would take between 250 and 500 one-MW wind turbines to provide the same effective load-carrying capability as a single 100-MW natural gas-fired plant. The lower effective load-carrying capability of wind turbines is one reason that a large-scale switch to wind power would be very expensive relative to producing an equal amount of power with more conventional power sources.

But the fact that individual wind turbines are, due to the intermittency of the wind, some of the least reliable of all power producers does not mean wind turbines cannot be used to increase system reliability. Probability theory shows that it is possible to construct a system that is more reliable than each of the individual power producers of which it is composed. By separating wind farms geographically, it becomes less likely that the wind will fail at all sites simultaneously. By building enough geographically separated wind farms, the probability that all will fail to produce power at the same time can be made as small as desired—provided, of course, that there are enough sites to accommodate the building boom. Constructing such a system would be extremely expensive; it would require a great deal of land, and a major and very expensive expansion of the required high-voltage transmission network. But provided one is willing to pay the price, wind power can make a significant contribution to system output and system reliability.

Storing the Wind

There is another way to overcome the problem of the intermittency of wind turbines. This method makes use of the process of electrolysis. Water molecules are comprised of one oxygen atom bound to two hydrogen atoms. In a process demonstrated in most high-school chemistry classes, one can split the molecular bonds that bind hydrogen atoms to oxygen atoms by passing an electric current through the water in a process called electrolysis. By collecting the resulting hydrogen gas, it is possible to store some of the kinetic energy of the wind as chemical energy. The hydrogen gas that results from the electrolysis process is highly flammable. When burned in a pure oxygen environment, the only product of combustion is water—or in a somewhat different process, the hydrogen can be used to power fuel cells, devices that convert chemical energy directly into electrical energy without employing the combustion process. The hydrogen produced by the turbine can be used on-site or collected and used elsewhere.

In theory, this process has a great deal of promise. Burning hydrogen produces no greenhouse gases and it is renewable in the sense that the hydrogen is obtained from water and produces water when it is burned. But electrolysis, as described here, has so far proved to be expensive and inefficient. There are three main reasons. First, hydrogen is very difficult to compress for storage. At low pressures, small amounts of hydrogen—where "small" is measured in terms of its energy content—require large storage containers and large amounts of hydrogen require huge storage containers. Alternatively, compressing hydrogen or liquefying it so that

WIND POWER, TOPOGRAPHY, AND THE ENVIRONMENT

As described previously, the maximum amount of wind energy that can be converted into electricity is very sensitive to the speed of the wind—more sensitive to wind speed than any other factor. This

it fits into smaller containers requires significant amounts of energy. Both the higher pressure and lower pressure alternatives are, therefore, costly. Second, electrolysis is not a very efficient process in the sense that a great deal more work is required to produce the hydrogen gas than is recovered by burning it. (Recall the discussion of the efficiency of heat engines in chapter 6.) And although fuel cells are, in theory, more efficient than heat engines, manufacturers have not yet found a way to produce fuel cells that are robust enough and economical enough for daily use in many common applications. Finally, because electricity is already a valuable product, a producer would have to be paid a premium for the hydrogen in order for it to be worthwhile to convert electricity that might otherwise be sold immediately on the grid into hydrogen gas that would later be used as fuel—especially given the inefficiencies of the additional conversion processes.

Engineers and scientists continue to investigate all the relevant technologies associated with wind power, hydrogen production and storage, and fuel cell manufacturing. All of these technologies are in a state of flux. It is possible that the difficulties presented by these technologies will be partially overcome, but it is also important to keep in mind that the promise of hydrogen as the "fuel of the future" dates back to the 1970s. Assertions that a new "hydrogen economy" is just over the horizon have been made every decade since the 1970s, and hydrogen supporters continue to insist that hydrogen will soon make significant contributions to the energy mix. Perhaps they are right.

raises two questions: Where are the best sites for development? And what are the environmental consequences associated with developing these sites?

From a technical standpoint, the "best" sites should have strong steady winds during periods of peak demand, and they should not

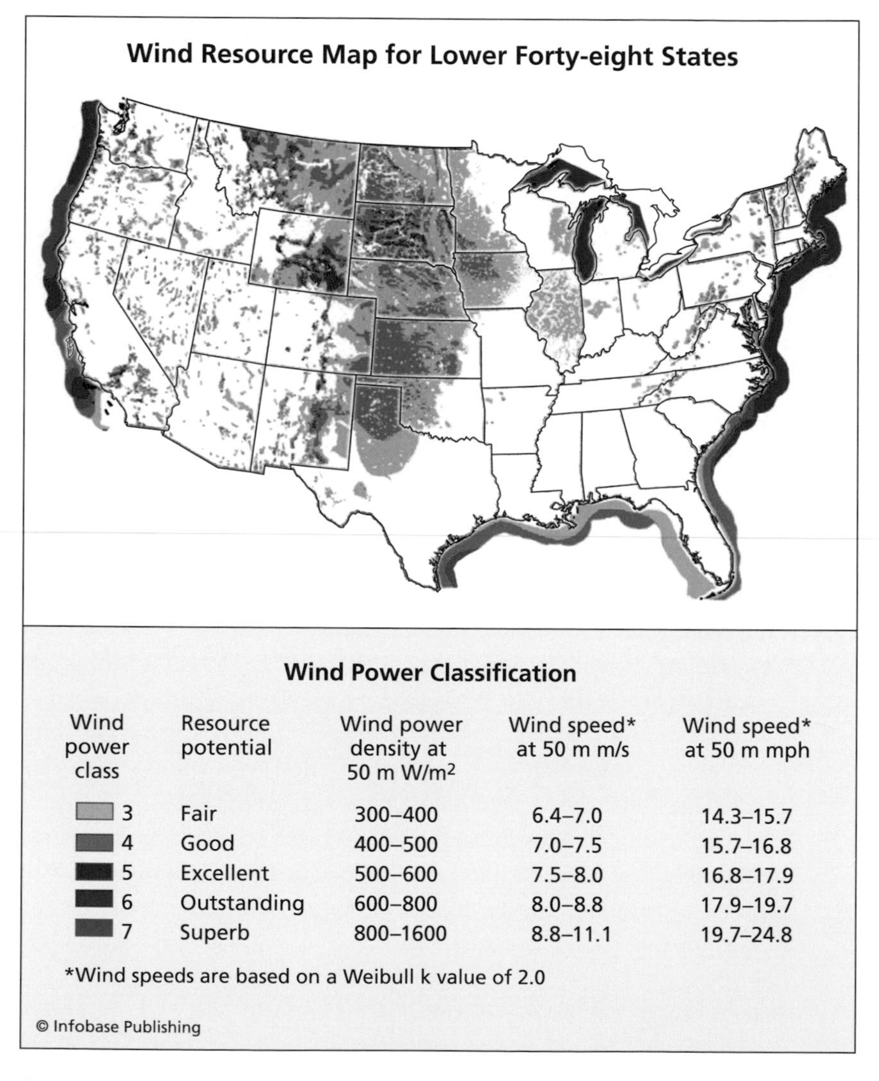

A wind resource map that shows the annual average wind power estimated at ca. 160 feet (ca. 50 m) above the surface of the United States *(Source: U.S. Department of Energy, NREL)*

be located too far from the markets that they are intended to serve. In the United States these sites have, for the most part, already been identified. The federal government has expended a good deal of money and effort creating what are called "wind resource maps," which, as the name implies, identify sites that could, in theory, be developed to provide significant amounts of electrical power. These maps show, for example, that the best sites occupy less than one percent of the land area in the lower 48 states—a useful fact, but that is only a beginning. Economically viable sites may be excluded from development because they are in national parks, scenic areas, wetlands, urban areas, or because the sites are home to endangered bird or bat species. Some sites are too isolated to make development attractive, and development of other sites is problematic because of fierce local opposition that is often based on reasons that are largely aesthetic. There is also the problem of scale: Commercial wind farms require many turbines distributed across substantial amounts of land.

Consider, first, the problem of building a wind farm on the crest of a mountain. The area along a mountain crest, also called a ridgeline, is often the best location for a wind turbine. In a mountainous area, ridgelines are sometimes the only viable locations because the valleys are sheltered from the wind by the mountains. Because one wind turbine is not economically viable, developers seek to place as many wind turbines as possible along a ridgeline. There are not many ways to build an economically viable wind farm on a ridgeline. The principal reason has to do with the wakes created by the rotors. As wind passes through the blades of a wind turbine some of the kinetic energy of the wind is converted into electricity. Downwind of the rotor, the wind is weakened. It is also turbulent, and ill-suited for driving another turbine. Consequently, those who design wind farms never position one turbine directly behind another. Along ridgelines, turbines must be positioned like beads on a string: single file and with space between them. Proposals to build such wind farms have sometimes run into vociferous opposition by those who, on aesthetic grounds, object to lining ridgelines with enormous wind turbines.

Alternatively, consider the development of wind farms in regions where the land is flat and the wind blows in one direction only. In this environment, the turbines could be organized into rows and the rows placed perpendicularly to the prevailing winds. Turbine spacing is expressed in terms of rotor length, and for an array of rotors on flat land with a steady wind, rows would ordinarily be separated by a distance of approximately 10 rotor diameters to allow the turbulence created by the upstream row the time and distance necessary to dissipate before encountering the next row of turbines. Within the same row, three to four rotor diameters between neighboring turbines is considered a reasonable distance. Because turbines have large rotors—230 feet (70 m) from tip to tip is not unusual—this means that rows would be separated by at least 2,300 feet (700 m, or about 0.5 mile), and within the same row, neighboring turbines would typically be from 690–920 feet (210–280 m) apart. Consequently, a wind farm consisting of three rows of 20 turbines would require at least two square miles (6.7 km^2) of land.

If the turbines in this model wind farm are rated at three MW each, the farm would have a maximum output of 180 MW and an effective load-carrying capability of between 36 and 72 MW—not a very large output, but spread out over a fairly large amount of land. Another way of describing the situation is that the wind farm could be expected to produce (on average) 28–56 kW per acre (69–138 kW/hectare). Evidently, wind energy on an industrial scale is land-intensive, and there is little to be done about it. Placing the rows much closer together risks making the turbines downwind of the leading row economically unviable—that is, the output from downwind turbines may not be enough to justify their construction if they are too close to the row upstream. Placing neighboring turbines much closer than about three rotor diameters has the same effect. By way of contrast, the Edwin I. Hatch Nuclear Plant site, which is located near Baxley, Georgia, and is home to two nuclear reactors, occupies 2,240 acres, most of which is used for timber production and wildlife habitat. The actual reactor site, which houses the reactor containment buildings, the cooling towers, and all structures associated with both reactors, occupies only about 300 acres. The two reactors on this site produce more than 1,700 MW of electricity or about 5,700 kW per acre (14,000 kW/hectare).

The preceding discussion may make it seem that one of the main disadvantages to large-scale wind development is that producing commercially valuable amounts of wind power requires large tracts of land. By itself, however, the requirement for large amounts of land does not necessarily lead to land-use problems, because wind turbines can sometimes share the landscape. Farmers, for example, have sometimes found it profitable to lease some of their land to power producers. The rotors spin high above the crops, and the amount of land required for the tower is relatively small. So the payments from the power producer to the farmer can more than offset the loss of income suffered by the farmer due to the relatively small amount of arable land taken out of production in order to make room for the turbines and associated infrastructure.

It is, of course, not just the size of the wind farm but its location that matters. As noted previously, the best locations occupy less than 1 percent of the landmass of the contiguous United States. (One can, of course, use sites where the wind is less powerful or less reliable, but building on less-productive sites also means building significantly more turbines to achieve the same power output. Producing power from wind energy at those less-desirable sites is less profitable and so less attractive as an investment opportunity.) Good sites for wind power are natural resources, as unique and irreplaceable as oil fields and coal mines.

10 Wind Energy: Economic and Public Policy Considerations

Electricity is a commodity, an article of commerce, and it can be produced by a wide variety of methods. Because wind energy is only one method among many for producing electricity, it must compete with other technologies in what many like to call "the" free market. There are, however, many free markets. Each nation creates its own legislation to establish the rules by which its free markets operate, and it establishes its own regulatory authorities to monitor compliance. There are some who insist on philosophical grounds that when it comes to free markets fewer rules are better, but this belief is not always supported by the available data. The actual situation is far more complex.

Today, the principal methods for generating electricity in the United States are coal, nuclear, and natural gas. If wind energy is to become a significant contributor of electrical energy, it will have to grow at the expense of one or more of these more conven-

U.S. Capitol. The federal government has been trying to create a wind-energy sector since the 1970s. *(Architect of the Capitol)*

tional sources of electrical power—that is, it is not enough for wind power to grow at the same rate at which the demand for electricity is growing. Rather, in order to displace other, more conventional power sources, wind power must grow faster than the demand for electricity. This is no easy task. A substantial infrastructure has evolved over the years to support coal, nuclear, and natural gas power plants. With respect to energy policies, each type of power-production technology has its own political constituency, and each fills a particular niche in the market. Natural gas plants, for example, make use of a highly developed production and distribution system to produce peak power (and sometimes base load power), and the base load producers, nuclear and coal, currently benefit from a highly developed infrastructure and substantial government subsidies of their own.

As described in chapter 7, the wind-power industry has also benefited from heavy government price subsidies and the benefits of government-funded research. But these subsidies are insufficient to guarantee that wind power will eventually displace significant amounts of more conventionally produced power. All power-generation technologies continue to develop and become more efficient as power producers struggle to either reduce their costs or to shift their costs onto others with the goal of becoming more competitive. If wind is to displace other power production technologies it must become more efficient than those other technologies or it must receive more in the way of subsidies. From a producer's point of view, of course, increased subsidies and increased efficiencies are not mutually exclusive goals.

THE COSTS OF WIND POWER

As has been demonstrated repeatedly in this volume, each power-generation technology has its own unique costs. There are environmental costs associated with every form of large-scale power generation, and there are the costs of financial subsidies used to encourage the further development and deployment of each technology. These costs are not always reflected in the prices charged to the consumer. In fact, they almost never explicitly appear in the monthly utility bill. They may, instead, appear in the form of higher taxes, increased environmental problems, or changes in government spending priorities—money may be shifted from health care to energy subsidies, for example.

There is, as has been previously mentioned, no generally agreed upon method of assigning value to particular environmental outcomes. It may be known, for example, that a particular power technology disrupts the environment in a particular way, but it is often less clear how to assign a cost to that disruption *relative to the costs associated with another type of power technology.* No modern nation can survive without access to large amounts of reasonably priced electrical power. Therefore, the value of any power technology can only be weighed relative to the available alternatives.

Financial subsidies are not always easy to recognize, either. In a budget as large and complex as that of the United States, for example, subsidies are often distributed among numerous agencies, and often they are not labeled as subsidies at all. In addition, higher taxes and spending shifts within the budget are practically always the result of numerous pressures, most of which have little to do with the energy industry. The costs of power production are real, but a full accounting of the costs associated with any method of power production has never been carried out—partly because of the difficulties just described and partly because the concept of cost is continually evolving.

To be sure, some of the subsidies that some power producers have received are related to campaign contributions, political patronage, and a host of other factors that sometimes gives politics a bad name, but which subsidies fall in the "undesirable" category is open to debate. As a general rule, energy subsidies simply reflect the fact that the provision of affordable and reliable electricity is an essential service upon which modern life depends. Electricity is a public good as well as a source of private wealth.

Some of the subsidies directed toward wind-power producers—the federal production tax credit and research conducted or at least sponsored by the Department of Energy, for example—were described in chapter 7. And some states—New York and California, for example—have historically also offered generous subsidies for wind-power producers. Today, U.S. wind-power producers benefit from substantial subsidies no matter where in the country they are located, although some localities still offer more financial support than others. Each year hundreds of millions of dollars in public monies are spent making wind power attractive to investors, who, without those subsidies, would likely spend their money elsewhere. Many of the taxpayers on whose behalf the subsidies are distributed do not use electricity produced by wind turbines.

And some wind-power costs will probably rise faster than the rate of installation of new wind facilities because the number of ideal sites is limited. As discussed in chapter 9, building a large number of turbines requires a great deal of land because turbines need to be placed far apart to prevent each turbine from interfering with the operation of its neighbors. Trees cannot be permitted to grow near turbines because they interfere with the flow of the wind. Turbines cannot be built near residential structures for two reasons: First, the buildings interfere with the wind just as the trees do, and second, turbines are noisy when in operation. Wind farms, therefore, require large amounts of inexpensive, sparsely populated, unforested land situated far from the nesting sites or migratory flight paths of rare or endangered bird and bat species. As the best sites are developed, secondary sites—that is, less-profitable sites—will be developed, and investors will require even higher subsidies to build in these locations. This is why it is reasonable to expect that subsidies will probably increase at a rate exceeding the rate at which new facilities are installed.

Wind farms can also be situated at sea, and a few offshore wind farms have already been built or are under construction in various nations around the world, but at present offshore projects are even more expensive than those on land and their successful operation requires even heavier subsidies. At present, it is not clear how much electricity from offshore wind farms consumers will be able to afford.

In addition to all of these restrictions, there is also the question of market access. Often wind farms are located far from the consumers who require electricity. By way of example, in New York (as of this writing) there is a good deal of controversy surrounding attempts to build a 200-mile (320-km) corridor for high-voltage power lines to connect upstate wind producers with downstate consumers. The proposal is controversial among those who will have to surrender some of their land for the corridor. From the point of view of the energy producers, however, the need for the project is obvious. Unless a way is found to connect their wind farms with the downstate energy markets, there is little point in developing further upstate capacity.

When all of these factors are taken into account, it is apparent that large-scale wind power costs a lot. But so do the alternatives. The energy requirements of modern societies are enormous, and meeting those requirements requires large-scale projects. If wind power becomes an important part of the energy mix—in the United States it currently supplies less than 2 percent of the total—it is reasonable to expect that the environmental and economic costs of wind will rise at least as fast as the output from the wind farms. There is no easy way to satisfy the energy requirements of modern societies.

THE ROLE OF ECONOMIC CLASS

The United States was once the leader in installed wind capacity, but despite a continuing commitment to wind power, it has not completely kept pace with developments elsewhere. Building a wind farm in the United States has become increasingly difficult. To see the problems faced by independent power producers as they attempt to increase wind capacity, consider the difficulties thus far encountered by Cape Wind, the company that hopes to develop the nation's first offshore wind farm. This wind farm will, if approved, be built just off the coast of Massachusetts on approximately 25 square miles (65 km^2) of Nantucket Sound.

Nantucket Sound is bounded by Cape Cod and the islands of Martha's Vineyard and Nantucket. Cape Cod and "The Islands" have some of the highest electricity prices in the nation, and in 2005 state regulators approved a 50 percent increase in electric rates for these consumers. About 60 percent of the electricity used by consumers in this area is produced by fossil fuel plants, and most of this is produced by burning expensive natural gas and oil. In addition to high-priced electricity and the accompanying

Brant Point Lighthouse on Nantucket Sound. If the Cape Wind project is built, it will be located miles from any land. *(Nelson Fontaine)*

greenhouse gas emissions, the power production capacity of the network has not kept pace with growing demand. ISO–New England, the organization responsible for ensuring sufficient generating capacity throughout New England, has consistently warned of impending shortfalls in electricity production. One would think, therefore, that construction of a 130-turbine emissions-free power project would be welcome in an area with an impending energy shortage, high energy costs, and a frequently expressed commitment to support clean energy, but this has not been the case.

Opponents to the project include radio personality Robert F. Kennedy, Jr., and his uncle, the late Senator Ted Kennedy, as well as the Alliance to Protect Nantucket Sound, which enjoys a multimillion dollar budget, former Massachusetts governor and past 2008 presidential candidate Mitt Romney, former Massachusetts attorney general Tom Reilly, U.S. Congressman William Delahunt, whose district includes Cape Cod, and other organizations and political figures. What are the objections?

During the course of the debate about the Cape Wind project, various environmental objections have been raised. By way of example, one early source of expressed concern was the effect of the turbines on birdlife. Wind turbines kill birds—this is well established—and when turbines are placed in areas with many birds, they will kill many birds. The key to minimizing the impact of a wind farm on avian life is to locate the turbines far from large populations of birds and even to locate them away from small populations of endangered birds. A thorough study of Nantucket Sound by independent scientists was undertaken in order to predict the probable impact of the project on birdlife. Scientists determined that most birds avoid Nantucket Sound, and radar studies have shown that although some migrating birds fly over the area, they fly at altitudes much greater than the proposed maximum height of the turbine blades. The impact on birdlife would, therefore, be small. Having established that this concern was without merit, objections simply shifted elsewhere.

Project opposition was, if not widespread, at least well financed. In 2004, the Alliance to Protect Nantucket Sound received $4.67 million in contributions, $2.9 million (or 62 percent) of which came from just 15 individuals. As each objection was raised, it was answered. Opponents, however, remained steadfastly opposed, although as their objections were answered they began to offer less and less in the way of explanation. In June 2006, for example, Senator Kennedy simply noted that, "There are still some outstanding questions with respect to the project," and Representative Delahunt noted that ". . . this project is destined for litigation that will be interminable." Neither was more specific. Opponents remained committed to preventing the project even after the 2007 announcement by Massachusetts Environmental Secretary Ian Bowles that the proposed Cape Wind project "adequately and properly complies" with state environmental requirements.

Of course, not everyone is opposed to the project. Voters, for instance, have for the most part not expressed their opposition at the ballot box. In the September 2006 Democratic primary for governor of Massachusetts, 53 percent of the voters on Cape Cod voted for Deval Patrick. Mr. Patrick was a supporter of the Cape Wind project. Christopher Gabrielli, another supporter of the Cape Wind project, was also a candidate in that primary. Together he and Mr. Patrick garnered 74 percent of the vote on Cape Cod, Martha's Vineyard, and Nantucket. Attorney General Tom Reilly, who was also running for the Democratic nomination for governor, and who vociferously opposed the Cape Wind project, ran a distant second to Mr. Patrick among voters on the Cape. In November of that year Mr. Patrick ran for governor in the general election against Republican Lt. Governor Kerry Healy. She was also firmly opposed to the Cape Wind project. Mr. Patrick won the general election.

What opponents to the project do not dwell upon is that the Kennedy family compound, which is located in Hyannis Port on Cape Cod, overlooks Nantucket Sound, as do many other expensive homes. The residents of these homes will see the wind turbines on the distant horizon when they look toward Nantucket Sound. Moreover, placing large wind turbines in the Sound may interfere with recreational boating. Perhaps for the first time, affluent consumers of electricity on Cape Cod are being asked to shoulder some of the burdens associated with electricity production—even if these burdens only affected their leisure activities and their ocean views.

While every large-scale power technology impacts the lives of many people, the burdens associated with production are never distributed equally. The residents of Appalachia have, for example, long borne much of the brunt of the United States' pro-coal policies. Mining activities have cost the lives of many mine workers and the health of many more; burning coal to produce power in the Midwest has had a profound effect on the forests and freshwater wildlife of the Northeast as pollution from Midwestern power plants has settled on the forests of New York and New England; transporting oil around the globe is a business that has long been punctuated by one environmental disaster after another; and oil refineries have adversely affected the health of some refinery workers. As a general rule those most able to afford the power produced by these technologies have been the least affected by the negative aspects associated with their use.

Historically, some groups, usually low-income groups or those identified as minority groups, have borne most of the environmental and health impacts of energy production technologies. This continues to be true. In response, those concerned with social justice have developed the concept of "environmental justice," the idea that policymakers should avoid crafting policies that disproportionately and adversely affect the health and environment of those communities least able to protect themselves. While everyone agrees that environmental injustices are unfortunate, not everyone agrees on a solution. Those advocating environmental justice assert that when the benefits of particular technologies are enjoyed by many, public policy should be crafted so that the consequences of implementing those technologies are more evenly distributed among the beneficiaries. As the experiences on Cape Cod demonstrate, the equitable sharing of the burdens associated with power generation and distribution is an ideal that is difficult to put into practice.

The lack of commitment to environmentally just policies is particularly evident in the development of wind power, because wind farms cannot always be sited among the poor and less powerful. Wind farms must be sited where the wind blows, and some of the sites best suited for wind power development are precisely those that have already attracted high-priced homes with views of mountain ridgelines or the open and relatively placid ocean vistas to be found in sounds and bays along the nation's coasts. Because wind farms require many large wind turbines, the visual impact of these projects cannot be entirely hidden from the view of those who live in these areas. Consequently, despite the fact that wind power is one of the most environmentally benign of all energy technologies, the construction of wind farms has sometimes proved to be highly controversial for reasons that are neither scientific nor economic. Simply put: Often the major objection to wind farms—and this is certainly the case for the Cape Wind project—is that those who will benefit from the power do not like the view.

THE FUTURE OF WIND POWER

How large a contribution can wind power make toward meeting the ever-growing demand for electricity in the United States and elsewhere? Although the question sounds simple, the answer is not. First, it is important to emphasize that wind power is only one technology among several. The goal of investors is not to produce power from wind but rather to produce power profitably. Profits can be the result of increased efficiencies or government subsidies, but without profits nothing will be built. Subsidies can flow directly to the wind-power industry as is the case with the production credits currently enjoyed by U.S. wind producers, or government support for wind can be indirect as would occur if the government were to tax power producers for emitting greenhouse gases, thereby creating a disincentive to future investment in fossil fuel plants, a policy adopted in Denmark. But while a carbon tax could benefit wind power, there is no guarantee that it will. Investments might be directed toward other technologies—tidal power, for example, or wave power—provided that tidal power or wave power are more attractive investments than wind power at the time that investors make their decisions. The effects of government policies on energy markets are complex, and historically, they are full of unintended consequences.

But the government, which establishes the legal regime in which energy markets operate, cannot retreat from its responsibilities in this area. One can only hope that legislators will approach the task with caution and humility, in part because there are many previous policies about which they have reason to be humble.

From a technical point of view, engineers have made numerous important improvements in wind technology. Since government researchers first became involved in the 1970s, the price of producing one kilowatt-hour of electricity from the wind has decreased by an order of magnitude—that is, it costs only about one-tenth as much to produce a kilowatt-hour of electricity from wind today as it did in the 1970s. The improvements have been incremental and largely the result of government-sponsored wind research. Small changes in blade designs, increasing the scale at which the machines are built, and other innovations have all had a cumulative effect, and the price of a kilowatt-hour of electricity produced by an "ordinary" wind turbine is still coming down.

Early wind projects were concerned primarily with developing land-based wind farms situated in areas with strong reliable winds, and while these remain important, emphasis has begun to shift to developing two other types of sites. First, engineers are increasingly looking toward the oceans. Offshore winds, free of the obstructions of mountains, trees, and buildings, tend to blow more predictably than over land. There are also fewer interests competing for the use of offshore sites than for land-based sites, and finally, as with the Cape Cod project, many offshore sites are located near the consumers who need the power. By contrast, many sites with strong winds that are located on land tend to be situated far from consumers.

Second, researchers are interested in developing regions where the wind speed is lower. The reason is that regions with lower but still commercially viable wind speeds are 20 times more common than regions with higher wind speeds. What one loses in output per turbine, one can gain by deploying more turbines. Of course, deploying multiple turbines simply to produce the same amount of power that a single turbine could produce in a region with higher and more reliable wind speeds is a recipe for higher-priced electricity. The goal is, then, to make these lower–wind speed turbines as cheaply and as efficiently as possible.

While wind enthusiasts claim that there is enough wind energy available within the United States and just off its coasts to meet a large fraction of the total electric needs of the United States, there are real questions about whether this energy can ever be exploited at a cost that consumers are willing or even able to pay. Furthermore, although the average amount of power that can be generated by the wind is large, consumers care little about averages. Instead, they want—they *need*—power at the instant that they turn on lights, computers, televisions, stoves, or heaters. In this sense, intermittence remains the central problem with wind power, but this is a problem that may have a solution as engineers find better ways to integrate wind power into the grid. This research is ongoing, and there are promising solutions under consideration. Time will tell if these solutions are practical.

Even if all technical problems are solved, wind power development may be stymied unless creative government policies can be developed that enfranchise those most likely to object to a particular project, the individuals living within sight of the wind farms. Such policies have already been crafted in Denmark. Perhaps similar policies will be adopted in other countries as well.

Currently, energy markets are in a state of flux as power producers seek to respond to increasing concerns about the cost of electricity and the environmental impacts associated with its production. New technologies, some of which have been described in previous chapters, may soon be competitive with wind. Some are, in fact, already competitive. While the long-term outlook for wind is difficult to predict, for the next several years it is likely that wind power will continue to grow at a rate that exceeds the rate of increase of electricity demand.

Conclusion

The technologies described in this volume can be divided into two classes: (1) conventional hydropower, and (2) everything else. Conventional hydropower is a mature technology. To be sure, engineers still make occasional improvements in turbine design, for example, and new ideas, such as pumped storage, are still occasionally introduced, but conventional hydropower is a mature technology in the sense that its limitations are well understood, and its potential has, for the most part, already been developed. All the other technologies in this book are still under development, and the potential of each of these methods for generating electricity is still a matter of some debate. Innovations in product design—as well as innovations in legislation that might lead to greater market penetration—are almost assuredly waiting to be discovered and implemented.

There are real advantages to developing these technologies. Emissions-free electricity is, of course, the principal advantage, but another important advantage is energy security. A disadvantage shared by all the technologies described in this volume except conventional hydropower is that the energy sources are all diffuse. There are, to be sure, enormous amounts of energy in the oceans' currents, Earth's winds, and the differences in temperature between the oceans' upper and lower layers, but there is not much energy at any given location. Harnessing diffuse sources of energy will always be expensive. By way of example, one can power a city with the output of a single nuclear reactor occupying a total of only a few hundred acres of land, but one would need to place thousands of wind turbines along many miles of coastline to accomplish the same goal. Even under these conditions the nuclear plant's output would be more reliable and more under the control of its operators than would the output of the many turbines that would be required to produce the same average power output as the nuclear plant.

As a practical matter, how much power is available from the technologies described in this book? The answer to this question is not yet clear—even in the case of wind, which is, with the exception of conventional hydropower, the most developed of all the technologies. Problems associated with intermittence of supply, local opposition, and the price of wind-generated electricity all continue to pose barriers to large-scale development. Subsidies help, of course, but while it is easy for any nation to afford the subsidies necessary to keep 100 turbines in operation, ensuring the profitability of 100,000 turbines would tax the budgets of every nation. As a general rule, as the number of turbines (and the subsidies they require) increase, public support can be expected to decrease. The future of any power-production technology is, in part, a matter of national priorities. Which, for example, is more deserving of support: wind energy, wave energy, space exploration, higher education, national defense, or health care? Each program has its proponents, but because even national budgets are finite, compromises must be made. It is unclear whether the power technologies described in this volume—with the exception of conventional hydropower—will ever produce more than a tiny fraction of the United States' power output. This is not to rule out the possibility of large-scale low-subsidy or subsidy-free electric power from these technologies, only to point out that this has yet to be accomplished.

Energy is one of the most fundamental problems of the 21st century. The ability to produce large amounts of electricity with minimal environmental disruption and at a reasonable cost is necessary if future generations are to build upon today's accomplishments. Each generation wants to see the next generation do better—a better standard of living, a better health-care system, better science, more peaceful relations between nations, an improved environment—but a necessary precondition for all of this is that solutions are found to the problems currently associated with energy production. Whether this will be accomplished remains to be seen.

Chronology

2900 B.C.E.	Egyptians build first known dam across the Nile
85 B.C.E.	Earliest reference to a watermill (in a poem by the Greek poet Antipater)
ca. 650	Vertical axis windmills in use in Persia
ca. 1100	Horizontal-axis windmills appear for the first time in Europe
1826	Lowell, Massachusetts, incorporated as a town. By 1850, it will be world famous as an industrial center dependent on entirely hydromechanical power
1827	First water turbine invented by the French inventor and engineer Benoit Fourneyron
1881	French scientist Jacques-Arsène d'Arsonval suggests the possibility of building what is now known as an ocean thermal energy converter (OTEC)
1889	Design for an impulse turbine is patented by the American engineer and inventor Lester Allen Pelton
1891	Danish inventor and teacher Poul la Cour begins to work on wind energy
1896	The hydroelectric facility at Niagara Falls begins transmitting power to Buffalo, New York, a major engineering achievement

1930 French scientist and engineer Georges Claude builds the world's first OTEC plant. It is deployed off the coast of Cuba

1941 Grand Coulee hydroelectric project completed

1970 Aswan High Dam completed in Egypt

1973 OPEC oil embargo

1978 The United States passes PURPA (Public Utility Regulatory Policies Act), the first concerted federal effort to encourage the construction of "alternative" energy sources

1979 U.S. funding for wind-energy research exceeds $50 million.

The Raccoon Mountain Pumped Storage plant, a project of the Tennessee Valley Authority, is brought online. At the time, it is the largest such facility in the world.

The first OTEC plant to produce usable amounts of energy (15 kW) goes into operation off the coast of Hawaii

1981 Consortium of Japanese companies put an OTEC plant into operation on the island nation of Nauru.

Denmark begins offering a production subsidy for wind power producers

1982 U.S. government funding for wind research is cut to $16.6 million

1983 The Itaipu hydroelectric power plant begins operation. (The last unit is brought online in 1991.) It supplies 78 percent of Paraguay's electricity and 25 percent of Brazil's electricity needs

1985 California achieves 1,000 MW of installed wind-generated power capacity

1989 U.S. Department of Energy funding for wind-power research reaches its lowest point since 1978

1990 The United States has 2,267 MW of installed wind-generated capacity.

Denmark has 300 MW of installed wind-generated capacity.

Germany has somewhat less than 100 MW of installed wind-generated capacity

1991 Germany passes the Electricity Feed Law guaranteeing wind producers a market for their higher-priced power

1992 The U.S. Energy Policy Act (EPAct) is passed, providing production credits to wind energy producers

1997 Germany surpasses the United States as the country with the largest wind energy capacity

2000 The German government passes the Renewable Energy Act with the goal of doubling the amount of electricity produced by renewable sources by 2010

2001 First Limpet (Land Installed Marine Power Energy Transmitter) begins operation on the isle of Islay off the coast of Scotland

2003 The Nathpa Jhakri hydroelectric power project, located in India and one of the more demanding civil-engineering projects of its era, begins operation

2004 First Archimedes Wave Swing pilot plant is connected to the grid.

Brazil produces 83 percent of its electricity from hydropower

2005 The U.S. Energy Policy Act of 2005 increases subsidies in wind power in order to make the technology more attractive to investors

2006 Europe installs 7.6 GW of wind power nameplate capacity in a single year for a cumulative wind power nameplate capacity of 48 GW.

The United States installs 2.45 GW of wind power nameplate capacity in 2006 for a cumulative wind power nameplate capacity of 11.6 GW

2007 The first commercial wave farm using Pelamis technology is installed off the coast of Portugal.

Scottish Power announces plans to install a Pelamis-type wave farm

List of Acronyms

AC	alternating current
AWS	Archimedes Wave Swing
DC	direct current
DOE	Department of Energy
EPACT	Energy Policy Act
FERC	Federal Energy Regulatory Commission
GW	gigawatt
kW	kilowatt
MW	megawatt
OPEC	Organization of Petroleum Exporting Countries
OTEC	ocean thermal energy converter
OWC	oscillating water column
PURPA	Public Utility Regulatory Policies Act of 1978
R & D	research and development
rpm	revolutions per minute

Glossary

alternating current an electrical current that reverses direction at regular intervals

amplitude in a wave, one half the distance from the peak to the trough

base load the minimum amount of electrical power delivered over a given period; the total amount of power minus transient increases in demand

capacity (1) actual power production divided by the theoretical maximum amount possible, also called the capacity factor; (2) the theoretical maximum rated output of power production, also called the nameplate capacity; (3) **effective load-carrying capability**

conventional hydroelectric power a power plant in which the electrical power is produced by a flowing stream that is usually regulated by a dam

cost (of power) the total value that must be surrendered in order to obtain a given amount of electrical power

direct current electrical current that flows in one direction only

effective load-carrying capability a measure of the contribution that a generating unit makes to grid reliability

efficiency the ratio of the energy supplied to a machine to the energy supplied by it

generator a device for converting mechanical energy into electrical energy

global warming the phenomenon marked by a gradual increase in average global temperatures caused by changes in the composition of Earth's atmosphere

greenhouse gases those gases that when added to the atmosphere in sufficient quantities cause **global warming**

hydraulic head the vertical distance between the surface water upstream and the surface water downstream of a hydroelectric power plant

hydraulic ram a type of pump that requires no external source of power other than the kinetic energy of the fluid flowing through it

impulse turbine a turbine commonly found in hydroelectric facilities with **hydraulic heads** in excess of 1,000 feet (1,600 m)

isoquant in a coordinate system where each point h on the positive part of the horizontal axis represents a value for the **hydraulic head** and each point q on the positive part of the vertical axis represents a volumetric flow rate (so that $p = hq$ is proportional to the power of a stream of water with hydraulic head h and volumetric flow rate q), an isoquant is the curve consisting of all points (h,q) such that hq equals a given value of p

kilowatt 1,000 watts

linear generator a device for converting linear motion directly into electricity

megawatt one million watts

nameplate capacity the maximum power output of a wind turbine under ideal conditions

peak power the amount of energy needed to meet electrical demand that is over and above the **base load** demand

penstock a conduit in a hydroelectric facility connecting the water supply to the turbine(s)

post mill a type of traditional European windmill in which the building that houses the windshaft and mill machinery rotates on a single post in order to face the wind

price the amount of money charged to a consumer for electricity, a quantity that often has little relationship to the electricity's actual cost

production credit a **subsidy** paid to a power producer, the amount of which is proportional to the amount of power produced

pumped storage facility a hydroelectric facility that generates power during periods of peak demand by releasing water that was pumped into an elevated reservoir during off-peak hours

reaction turbine a **turbine** designed for a hydroelectric facility with a medium or low (less than 600 feet [180 m]) **hydraulic head**

rotor (1) that part of a wind turbine consisting of the blades and the hub on which they are mounted (2) that part of a generator that rotates within the stationary cylindrical device called the stator

sail in a windmill, the large surface consisting of a (usually) wooden lattice and a canvas or wooden covering. It transmits the force of the wind to the **windshaft**

step-down transformer a device used to reduce the **voltage** of an **alternating current**

step-up transformer a device used to increase the **voltage** of an **alternating current**

subsidy a grant by a government designed to facilitate research into, development of, deployment of, or operation of a particular power-generation technology

tidal barrage a hydroelectric facility that depends on the rise and fall of the tides in order to generate the necessary **hydraulic head**

tidal mill a device conceptually similar to a wind **turbine** that converts the energy of flowing waters generated by the tides into electricity

tower mill a type of traditional European windmill. It is built in the shape of a truncated cone, and the windshaft is housed in a moveable cap

transformer a device used to change the voltage of alternating current

turbine a device used to convert the linear motion of a moving liquid or gas into rotary motion

voltage a measure of the difference in electrical potential, it is sometimes called the *electrical pressure* in analogy with water pressure

windshaft in a windmill, the shaft to which the **sails** are attached; in a wind **turbine,** the shaft to which the **rotor** is attached

Further Resources

The means by which nations produce their electrical power has become a very controversial topic. As this book indicates, there are real physical constraints on every type of power-production technology, and the value of any particular technology is determined, in part, by the nature of these constraints. The better these limitations are understood, the more accurately the value of each technology can be assessed. The following books and Web sites emphasize the science of power production.

Berinstein, Paula. *Alternative Energy: Facts, Statistics, and Issues.* Westport, Conn.: Oryx Press, 2001. The author has a tendency to try to sell each type of energy described, but the chapters contain many interesting facts not easily found elsewhere.

Boyle, Godfrey. *Renewable Energy: Power for a Sustainable Future.* 2nd ed. Oxford: Oxford University Press, 2004. Highly recommended. This textbook is almost 500 pages long. Readers of the present volume will be well-prepared to tackle those sections of Boyle's work that concern wind and waterpower. *Renewable Energy* is as advanced and complete a treatment of this important topic as one can find that does not use calculus.

Hay, Duncan. *Hydroelectric Development in the United States, 1880–1940.* Washington, D.C.: Edison Electric Institute, 1991. Most of the big projects in the United States were built or at least under construction by 1940. This book conveys something of the impact that these projects had on the development of the United States as well as some of the design considerations that went into the projects themselves.

Hostetter, Martha, ed. *Energy Policy.* Reference Shelf, vol. 74, no. 2. New York: H.W. Wilson, 2002. A collection of articles, reprinted from reputable magazines and newspapers, discussing various aspects of energy and energy policy. Very interesting.

Houghton, John. *Global Warming: The Complete Briefing.* 3rd ed. Cambridge: Cambridge University Press, 2004. Global warming is the topic that drives much of the interest in so-called alternative forms of energy. *Global Warming* offers a thoughtful and reasonably thorough description of what is presently known about the phenomenon of human-induced climate change.

Hunt, V. Daniel. *Windpower: A Handbook on Wind Energy Conversion Systems.* New York: Van Nostrand Reinhold, 1981. During the late 1970s and early 1980s, a number of high-quality books about wind energy aimed at a general readership were published. This is one of them. While the technology has changed a lot in the intervening years, wind has not. This book is still a first-rate source of information on the physics of wind and the principles involved in the conversion of wind energy to electrical energy.

National Research Council, Committee on Nuclear and Alternative Energy Systems. *Energy in Transition, 1985–2010: Final Report of the Committee on Nuclear and Alternative Systems, National Research Council, National Academy of Sciences.* San Francisco: W.H. Freeman, 1980. Highly recommended to all those who enjoy reading books that make predictions about the future of energy research and development. This almost-700-page book was written by some of the best engineers and scientists in the United States at the time, and they got almost everything wrong. For example, only four pages of this tome are devoted to wind energy, but whole chapters are devoted to breeder reactors, controlled nuclear fusion, and solar energy. It turns out that in the United States wind energy now produces more power than all three of those other sources combined.

Paton, W.R., trans. *The Greek Anthology*. Vol. 3. New York: G.P. Putnam, 1918. This book is the source of the description of the ancient watermill found in chapter one.

Pool, Robert. *Beyond Engineering: How Society Shapes Technology*. New York: Oxford University Press. 1997. While not about wind and waterpower specifically, this book addresses the interplay between society and technology and how the evolution of each affects the other. This subject is of vital interest to those who want to change the way power is generated.

Shaw, Jane S., and Manuel Nikel-Zueger. *Energy*. Critical Thinking about Environmental Issues. Farmington Hills, Mich.: Greenhaven Press, 2004. A general discussion about energy and the environment. Elementary but well-written.

Williams, Wendy, and Robert Whitcomb. *Cape Wind: Money Celebrity, Class, Politics, and the Battle for Our Energy Future*. New York: Public Affairs, 2007. A revealing report on the difficulties involved in obtaining approval to build the United States's first offshore wind farm.

INTERNET RESOURCES

Energy production has become a very controversial subject. Each technology has its proponents, and it is not always easy to find an evenhanded description of how a particular method of power generation works. As you read, be skeptical!

Danish Wind Industry Association. "Wind Turbines Deflect the Wind." Available online. URL: http://www.windpower.org/en/tour/wres/tube.htm. Accessed August 13, 2008. As a result of the pioneering work of Danish scientist and inventor Poul la Cour, Denmark has been involved in the development of wind-generated electrical power right from the beginning. This informative site gives an excellent overview of the basics surrounding wind power.

Elliot, D.L., C.G. Holladay, W.R. Barchet, H.P. Foote, and W.F. Sandusky. "Wind Energy Resource Atlas of the United States." Available online. URL: http://rredc.nrel.gov/wind/pubs/atlas/. Accessed August 13, 2008. Prepared for the U.S. Department of Energy, this document discusses the basics of wind, the raw material on which wind turbines depend. There are wind resource maps as well as explanations that convey the technical details of how wind energy is quantified.

Energy Information Administration. Official Energy Statistics from the U.S. Government. Available online. URL: http://www.eia.doe.gov/. Accessed on August 13, 2008. This is certainly the best source of statistics with respect to how energy is used. It has separate sections on all major technologies and most minor technologies. The Web site is not especially easy to search, but it is well worth the effort.

Gompertz, Simon, Business Correspondent, BBC. "On the Brink of a Wave Revolution." Available online. URL: http://news.bbc.co.uk/2/hi/programmes/working_lunch/4849540.stm. Accessed August 13, 2008. At the bottom of the page are links to BBC videos that show the Pelamis wave energy converter and a tidal mill in action. The videos convey something of the size of these projects as well as the difficult environments in which these machines function.

Kotchen, Matthew J., Michael R. Moore, Frank Lupi, and Edward S. Rutherford. "Environmental Constraints on Hydropower: An Ex Post Benefit-Cost Analysis of Dam Relicensing in Michigan." Available online. URL: http://www.msu.edu/user/lupi/Kotchen_etal_InPress_LandEcon2006.pdf. Accessed August 13, 2008. This is the paper described in chapter 3. It is not particularly easy reading, but it contains information on how federal legislation, economics, and physics interact to affect the way that hydroelectric facilities are used.

National Renewable Energy Laboratory. "What Is Ocean Thermal Energy Conversion?". Available online. URL: http://www.nrel.gov/otec/what.html. Accessed August 13, 2008. This site provides an overview of OTEC technology as well as a map that shows where the oceanic temperature differences are best suited for deploying OTEC technology.

U.S. Bureau of Reclamation. Columbia Basin Project. Available online. URL: http//www.usbr.gov/dataweb/projects/Washington/columbiabasin/history.html#Construction. Accessed August 13, 2008. This is an excellent and very thorough history of the Grand Coulee Dam, one of the largest civil-engineering projects in the history of the United States.

U.S. Department of Energy. "Wind Power Today." Available online. URL: http://www.nrel.gov/docs/fy06osti/39479.pdf. Accessed August 13, 2008. Although this presentation is a little one-sided, it describes a number of important wind-power research projects and the potential contribution of wind energy to United States energy markets.

U.S. National Park Service. "Building America's Industrial Revolution: The Boott Cotton Mills of Lowell, Massachusetts." Available online. URL: http://www.nps.gov/nr/twhp/wwwlps/lessons/21boott/21boott.htm. Accessed August 13, 2008. Designed for teachers, this excellent site has a number of photographs and drawings in addition to the well-written text. It tells an important part of the story of the industrialization of the United States.

Index

Note: *Italic* page numbers indicate illustrations and diagrams; page numbers followed by *c* indicate chronology entries.

D

F

M

N

R

X